GREEN PLASTICS

GREEN PLASTICS

An Introduction to the New Science of Biodegradable Plastics

E. S. Stevens

PRINCETON UNIVERSITY PRESS · PRINCETON AND OXFORD

Published by Princeton University Press, 41 William Street, Princeton,
New Jersey 08540
In the United Kingdom: Princeton University Press, 3 Market Place, Woodstock,
Oxfordshire OX20 1SY

Library of Congress Cataloging-in-Publication Data

Stevens, E. S. (Eugene S.), 1938-
Green plastics : an introduction to the new science of biodegradable plastics /
E. S. Stevens.
p. cm.
Includes bibliographical references and indexes.
ISBN 0-691-04967-X (acid-free paper)
1. Biodegradable plastics. I. Title.

TP1180.B55 S74 2002
668.4—dc21 2001036257

British Library Cataloging-in-Publication Data is available

This book has been composed in Times New Roman

The publisher would like to acknowledge the author of this volume for providing the
camera-ready copy from which this book was printed

Printed on acid-free paper

www.pup.princeton.edu

Printed in the United States of America

10 9 8 7 6 5 4 3

Contents

Preface

At the beginning of the twenty-first century, plastics are a leading material, providing uncountable useful and inexpensive items for modern living. This book brings attention to the emergence of **bioplastics**, a new generation of biodegradable plastics whose components are derived entirely or almost entirely from renewable raw materials. They conserve irreplaceable fossil fuels, contribute little to the already burdensome problems of waste management, and help prepare us for the time when fossil fuels become exhausted.

The word bioplastics has sometimes been used to mean biodegradable plastics, whatever their origin. It has also been applied to biomedical plastics, whatever their origin or biodegradability. The present usage associates the word bioplastics with renewable origin as well as with biodegradability, in order to capture for these intriguing new materials both aspects of their extreme naturalness.

Because the point of departure here is the environmental impact of plastics, the emphasis is on commodity materials—those produced, consumed, and discarded in large amounts. Specialty biomedical materials, of high value and high cost, have negligible environmental impact and are only mentioned briefly.

Time will tell whether bioplastics can secure a place in the current Age of Plastics. Their widespread use depends on developing technologies for successful commercial production, which in turn will partly depend on how strongly society is committed to the concepts of resource conservation, environmental preservation, and sustainable technologies. But there are growing signs that people indeed want to live in greater harmony with nature and leave future generations a healthy planet. The development of bioplastics technologies will also have the effect of creating new markets for agricultural products, perhaps revitalizing languishing rural areas.

The aim of the book is to introduce these earth-friendly—but as yet little known—plastics to a wide audience, including readers with only a limited knowledge of chemistry. Professionals not working in the field of biodegradable plastics but interested in learning the basics will find it a concise introduction to this exciting interdisciplinary topic.

Part One describes the growing concern over the environmental effects of using plastics in ever-increasing amounts. It also introduces the chemical nature of synthetic polymers and gives a brief account of the environmental degradability of plastics.

Part Two describes naturally occurring polymers as alternative plastics feedstocks, contrasting them with synthetic polymers. It also introduces the emerging bioplastics technologies. The book ends with directions for making cast-film samples of bioplastics using nothing more than commonly available items. The preparations vividly illustrate the potential of bioplastics, and are easily adapted for use in instructional laboratories.

Text enclosed in boxes provides more detailed descriptions of chemical terms and structures than may be required by the general reader. A Notes section contains individual citations; those references are only a small sampling of the growing literature on the subject. The Notes also include references to additional material for readers who want more detailed information on specific topics. A Glossary provides many definitions; important terms appear in boldface, usually when first used in the text. The Reading List includes additional resources.

The field is changing rapidly, both with respect to advances in research and technology, and with respect to the commercial ventures that aim to translate technical knowledge into useful products. I have attempted to make the book as up-to-date as possible.

E. S. Stevens

Acknowledgments

Many scientists have contributed to the field of degradable plastics made from renewable resources. I am indebted to all and ask pardon that the references are more illustrative than comprehensive.

In the early stages of writing, colleagues at Binghamton University provided stimulating words of encouragement: Janice Musfeldt, Bruce Norcross, Richard Quest, and Eric Stroyan of the Chemistry Department; Burrell Montz and Herman Roberson of the Department of Geological Sciences and Environmental Studies. Gregory T. Stevens, my son, made many suggestions that helped me keep younger generations of readers in mind.

Randal L. Shogren, of the National Center for Agricultural Utilization Research, U. S. Department of Agriculture, Peoria, Illinois, and Graham Swift, of G. S. Polymer Consultants, Blue Bell, Pennsylvania (recently retired from Rohm and Haas Company, Spring House, Pennsylvania), well-known experts in the field of degradable polymers, read early drafts and made valuable comments. I am grateful for their interest and time. In the end I alone am responsible for the views expressed and any omissions or inaccuracies that remain. My editors at Princeton University Press provided supportive advice and helped me see the project through to completion.

Several people helped locate, select, and provide permissions for photographs: Alan Miller, Seattle Art Museum; Jon B. Eklund, National Museum of American History, Smithsonian Institution; Mark Abbott, London Science Museum; Keith Lauer, National Plastic Center and Museum, Leominster, Massachusetts; Mrs. Julie Robinson; Research Center Staff, Ford Museum, Dearborn, Michigan; Randal L. Shogren, U.S. Department of Agriculture; and Richard P. Wool, Center for Composite Materials, University of Delaware. The photographer for Figures A.1–A.3 and A.6 was Christopher Focht.

PART ONE

PLASTICS

1

The Age of Plastics

The New Kids on the Block

No material on earth has been so highly valued for its usefulness, yet so maligned, as **plastic**. We have ambivalent, contrary, and vacillating feelings about plastics, and have never finally decided whether plastics are the good, the bad, or the ugly. One reason for the ambivalence is probably their newness. The rapid growth of plastics production was a twentieth-century phenomenon, and anything less than a hundred years old, on a historical scale, is novel. Among materials, plastics are newcomers, and we simply have not had time to make up our minds about them.

Plastics are so clearly useful that it is foolish not to afford them major respect. They are often not only *less expensive* than alternative materials, but their properties often make them *better*. Their low cost has undoubtedly had life-saving consequences, as in drought-prone areas of Africa where lightweight plastic water pails, at times the most important family possession, have replaced clay and stone containers, making it possible to bring in water from even distant wells in times of severe water shortage. Plastics are also perfectly matched with the modern information-age uses of cell phones, bank cards, and laptops. And even when mere comfort is at stake, no one can deny plastics are outstanding performers. Synthetic fibers, cousins to plastics, have become so highly developed that even the most die-hard naturalists turn to them to keep warm and dry working out-of-doors on a damp winter's day, or simply working up a sweat on a crisp cool ski slope.

But plastics, being so inexpensive, run counter to the usual association of good with rare and costly—the snob-appeal factor. Gold is good; silk and satin are good; but what are we to make of plastics, which *anyone* can own?

Their low cost and versatility have also allowed an unprecedented range of applications. In a free market all market niches tend to get filled, so that plastics have taken on every imaginable form. People's tastes being as varied as they are, there are differences of opinion on the aesthetic value of some of them. What one person finds fetching, another finds garish—and the material is condemned along with the form given it. The fact that many beautifully designed plastic objects are manufactured has never seemed to provide enough weight to balance the view that associates plastics with aesthetic poverty. Plastics may never shed the guilt-by-association burden, because their low manufacturing cost will always allow the mass production of objects of disputable beauty.

Moreover, the synthetic nature of plastics has come to stand for artificial and not-genuine, with connotations of phony or false. (He is so *plastic*!) The combined effect of tawdry applications and conflation of synthetic and false has been to color the popular attitude toward plastics.

Some singular voices have even been raised connecting plastics with all that is bad in society—a "malignant force" set loose to wreak havoc. But, in the remarkable breadth of human opinion, countervoices have unstintingly and exuberantly sung their praise. Nylon is not only practical, it's *sexy*. Vinyl phonograph records produce the only truly *authentic* sound. Andy Warhol wanted "to *be* plastic."

This book is not about the sociology of plastics, and it is not about the role that plastics play, or do not play, as the cause of, or the reflection of, deep-rooted social, political, cultural, or economic truths. It tells the story of the recent, as yet tentative, emergence of new plastics with characteristics not usually associated with plastic—plastics made from natural, renewable starting materials, plastics that are able to biodegrade totally and completely in an environmentally benign manner.

Not being a plastics industry insider, I am not privy to the long-term plans being worked out in the board rooms of the plastics industry. It is possible that even the movers and shakers of the plastics industry do not know exactly what paths the industry will be taking five or ten years from now. But as a chemist with a thirty-year professional relationship with molecules, particularly the large polymer molecules found in nature, I see these natural polymers coming into their own as starting materials for a new breed of plastics.

If these new plastics make their mark, it will be a comeback, a revival, rather than a totally new appearance, for they have not been completely unknown in the past. Perhaps the closest we have ever come to having a major presence of plastics made from natural polymers was when Henry Ford began a substantial research project aimed at making plastic automobile parts out of soybeans! But his plan was cut short by World War II. Had the soybean venture worked out, we might have had by now a new slang expression, "—or I'll eat my car."

The starting point in the story of these new **bioplastics** is the simple fact that plastics are now so commonplace that they have become an integral part of everyday life. There are personal use items, like the toothbrush, comb, ballpoint pen, and credit card. There are containers, like the jug of milk and the bag that holds the loaf of bread. And there are the wrappings on all those articles we purchase, like drugstore items, clothing, and videocassettes.

Plastic comes in all sizes and shapes. It can be molded, like the comb and toothbrush, or formed into sheeting or films. Some items are only partly made of plastic; others are made entirely of plastic, but of more than one type of plastic, fabricated to make a useful item.

Production of plastics on a very large scale is relatively new. The Dustin Hoffman character in the 1967 movie *The Graduate* was advised to go into "Plastics!" if he wanted a promising career and a prosperous future. That future is now. In the United States plastics industry over 20,000 facilities produce or distribute raw materials, molds, processing machinery, or products. They employ over one and a half million workers and ship more than $300 billion in products annually.

Past ages of human society have been called the Stone, Bronze, Copper, Iron, and Steel Ages, according to the material most used to fabricate objects. Today the total volume of plastics produced worldwide has surpassed that of steel and continues to increase.

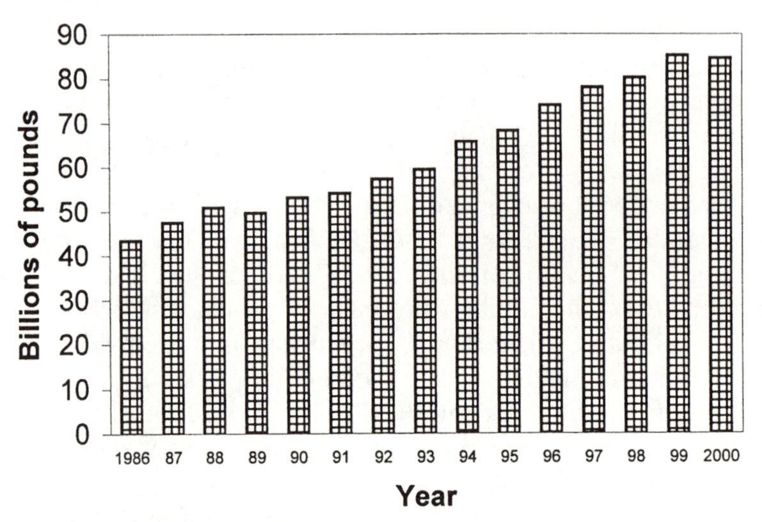

Figure 1.1 *Plastics production in the United States*

Approximately 200 **billion** pounds (100 million **tons**) of plastics are produced each year, with over 80 billion pounds a year being produced in the United States alone (fig. 1.1). We have entered the Age of Plastics.

How Do We Use All That Plastic?

The phenomenal rise in the use of plastics is the result of their extraordinary versatility and low cost. They make a good match with the needs of our rapidly growing world population. But if 200 billion pounds of plastics are produced each year, that's about 40 pounds a year for every person on the planet. What do we do with it all?

Much of the plastic that is produced is used for packaging. In the United States, about 30 percent of the plastic produced each year, over 20 billion pounds, is used for packaging, representing its largest use by far (fig. 1.2). In Western Europe 42 percent of all plastics use is for packaging.

Many people remember items that were previously sold un-packaged in bins but are now packaged individually or in groups of

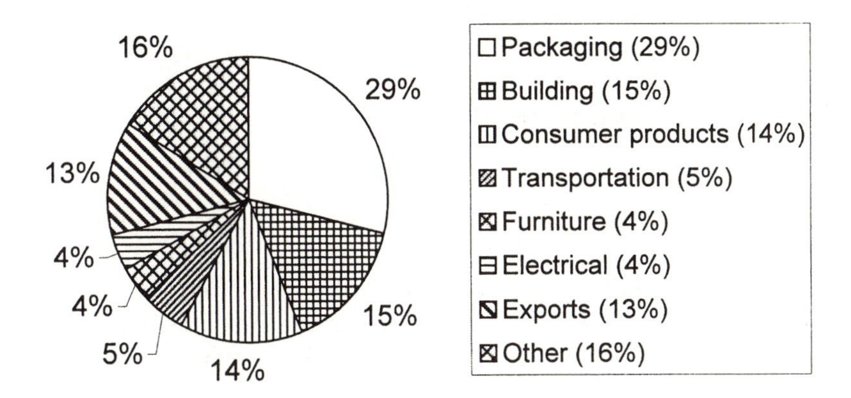

Figure 1.2 *Uses of plastics produced in the United States*

some small, or large, number. The purpose may be to provide added protection, longer freshness, or some other benefit to the consumer; it may be for inventory or some other purpose of the seller.

Plastic packaging is popular on account of its low cost and performance properties. There are now many forms of it, from plastic shopping bags to different types of plastic loose-fill packaging material, including the peanut-shaped variety.

Approximately one-half of the plastic used in packaging is for containers, such as soft-drink bottles and jugs for milk, water, laundry detergent, and bleach. One-third is in the form of plastic sheeting or film for items like bread wrap and grocery sacks. The remainder is for closures (caps and bottle tops), coatings, and other purposes.

Both flexible plastic packaging and semirigid plastics have been growing in use in *food* packaging so that now, although paper and paperboard packaging still dominates, plastic food packaging has become second in importance, followed by metal, glass, and other materials. Food and beverage packaging accounts for approximately 70 percent of the more than $100-billion packaging market in the United States and more than half the $400-billion worldwide market.

The popularity of microwave ovens has contributed to the rapid growth of plastic food packaging because they require the use of nonmetal containers. Many plastic packages are now designed to go conveniently from the freezer to the microwave oven, to the dinner table, and then directly to the trash bin.

As plastic packaging has increased, the use of synthetic packaging adhesives has also grown, in order to maintain compatibility. Plastic surfaces are often difficult to bond, the packaging is frequently very flexible, and the processing of packaging materials is typically rapid. Natural adhesive materials made from starch, dextrin, and sodium silicates, although cheaper, have not been able to compete in some packaging markets, and there has been a large increase in the production of synthetic packaging adhesives. Over a billion pounds of synthetic adhesives are used in the United States each year for packaging. We have become a plastics-oriented society partly because we have become a packaging-oriented society.

But plastics are versatile and are used for much more than packaging. *Building materials* of heavy-duty plastic, often replacing metal and wood, are manufactured in the United States to the extent of nearly 20 billion pounds a year. *Consumer products* include eating utensils, toys, diaper backings, cameras, watches, sporting goods, personal-hygiene articles like combs and razor handles, and much more. Institutional use of some of these items, like plastic eating utensils in schools and hospitals, makes plastics use for consumer products very large. In the United States over 10 billion pounds of plastic are turned into consumer products each year.

Transportation uses for automobile, watercraft, and aircraft parts total more than 4 billion pounds a year. *Furniture* accounts for almost 4 billion pounds a year. *Electrical components*, including wiring insulation, are commonly plastic.

There are many other miscellaneous end uses of plastics, each accounting for a billion pounds or so a year or less. Plastics are used on a large scale for trash bags, which might be called packaging for trash. Around a billion pounds of plastics are manufactured each year for that purpose alone. Industrial plastic sheeting is also used widely. In manufacturing industries there are many plastic machinery components. Plastic materials are used as coatings for paper and cardboard. Hospital equipment, scientific-research equipment, and military equipment all have plastic components.

Agricultural uses of plastics are more important than the scale of production indicates. Plastic ground covers, for example, are used to increase crop yield by as much as 200 or 300 percent. Just as home gardeners use mulch to conserve moisture, raise the soil temperature, prevent nutrient loss, and inhibit proliferation of weeds and insects, farmers use plastic agricultural covers for the same purposes but on a much larger scale. The use of agricultural covers is driven by economics. If the increase in crop yield outweighs the costs of producing and using the cover, the cover has an advantage. The use of plastic agricultural covers on a scale of millions of acres is important for increasing food production for a growing world population, and it is significant in terms of the vast amount of plastics required.

Other large-scale agricultural uses of plastics are for plant containers, binders and twines, irrigation products, netting to protect crops from birds, and temporary covers for storing grain.

There is a constantly growing number of uses for plastics in *biomedical* applications. They vary from the ordinary, like gloves, masks, gowns, plastic wraps, and coverings, to the more complicated, such as sutures, other wound-closure products, and drug-delivery systems, to the extraordinary, including orthopedic-repair products and other implants. Plastics used for the more complex biomedical applications are expensive and are not produced on the same large scale as the high-volume, low-cost commodity plastics that account for most of the use of plastics.

The use of plastics has grown so remarkably because of the large number of applications that have been developed for them. Plastics have become an important part of modern life and are here to stay. They have, however, raised the question of reconciling convenient living with concern for ecology.

2

Plastics as Materials

Materials Science

A current keyword in science is *materials,* as in *materials science*. It is a new enough term not to be included in many textbooks, but the concept behind it has come to be a major driving force. The word **material** refers to any matter that can be fabricated into useful products. Wood, metals, glass, ceramics, and plastics are all examples of materials. (Here the word *material* is used in the general, descriptive sense, not in the technical sense of a composition of matter complying with defined standards.)

Materials science is the study of compositions of matter that have application in the fabrication of useful products. In materials science the potential usefulness of the studied materials is never far from mind, and that emphasis on practicality has given materials science its own identity as a special branch of science.

Materials science is interdisciplinary; it includes aspects of chemistry, engineering, physics, biology, and geology. A materials chemist formulates new materials and characterizes them chemically; an engineer or physicist specializing in materials might measure the physical properties of newly formulated materials, or develop processes for manufacturing materials so that they possess particular properties. If part of the material derives from biological sources, biologists work to optimize properties, sometimes through genetic engineering. Geologists bring their specialized knowledge of minerals and ceramics to bear on the material of interest. But apart from its specializations, materials science is the study of the material as a whole, and project chemists, engineers, physicists, biologists, and geologists work closely together.

Materials science is growing. Colleges and universities now have courses and degree programs in it, and the number of materials scientists will undoubtedly increase. Governments and industries are supporting the field in ways that recognize its practical importance. In the United States, the National Science Foundation has identified materials science as a high-priority area in its allocation of federal tax dollars.

One important focus in materials science, even before that field took on its present name, was on making materials *strong*—for increased reliability, greater permanence (and therefore lower cost), and wider range of application. More recently a broader view has been adopted when asking what makes materials *better,* and there has been a surge in the number of directions research is taking. Molecular self-assembly processes are being exploited to produce particles with well-defined and reproducible structures in the *nanometer* (10^{-9} meter) range, giving rise to new nanoparticle-based materials and the burgeoning field of *nanoscale technology*. Thin membranes made from a variety of materials (including plastics) are being incorporated into devices that are specifically permeable to one or another type of molecule. Electrically conducting materials are being regularly fabricated without the use of metals. Microsensors continue to shrink in size, even as they grow in sensitivity and reliability. Increasing strength is still often a goal; novel ceramic materials are reaching new levels of hardness and strength.

The current trends in materials science are based on the premise that designing materials with particular *macroscopic* properties requires an understanding of materials at the *molecular* level. So chemists, who take satisfaction in thinking of chemistry, the study of molecules, as the "central science," have support from materials science for their view.

Plastics research is thriving along with the other branches of materials science research, and is more diverse than ever. As with other materials, increasing strength is still the goal for some applications. Polyethylene can now be engineered to be strong enough for use in artificial knees and hips. Fibers of oriented polyethylene can be made so strong that they serve as reliable tethers in space missions. But horizons have expanded, and plastics research now encompasses a wider range of goals—like plastics with specific permeability or barrier properties, or plastics with particular electrical properties. Some plastics research even encompasses an objective contrary to permanence—plastics designed to degrade according to specific timetables.

Composites and Laminates

Within the expanding boundaries of materials science research, one area of rapid growth has been the search for new composites. A **composite** is a solid product consisting of two or more distinct phases, including a binding material (matrix) and a fibrous or particulate material. Examples are plastics with fiber, metal, or mineral reinforcements. In fiber-reinforced plastics the matrix serves mainly to bind the fibers together; high-strength composites require strong interface bonds. But the matrix is also tailored so that the final composite has specific strength, impact resistance, and other properties.

In the construction industry, fiber-reinforced plastics containing as much as 90 percent wood fibers and 10 percent binding matrix are used for particle board and other construction products. High-strength *advanced* composites are made with up to 70 or 80 percent glass or graphite carbon fibers. Glass-reinforced plastics can be stronger than steel, and their density of 1.5 to 2.2 grams per cubic centimeter (g/cm^3) is less than that of steel (around 7.9 g/cm^3) or even aluminum (2.7 g/cm^3). Automobile body panels, truck and farm machinery panels, aircraft and military equipment, and speedboat hulls are commonly made of sturdy glass-fiber-reinforced composites.

Graphite carbon fibers are made by carbonizing fibers of a polymer—such as polyacrylonitrile—at temperatures of 1000 °C, producing fibers of 90 percent carbon. The carbonization level is then raised to as high as 99 percent by heating at temperatures of several thousand degrees. Carbon-reinforced composites can replace metal parts because of their strength—and they do not corrode. They are used in everything from aircraft parts and structural components on bridges to windmill blades and sporting goods, including tennis racquets, fishing rods, golf clubs, and skis.

A broader definition of composite includes not only heterogeneous materials but also homogeneous *blends,* including novel blends of plastics.

A **laminate** is a product made by bonding together two or more layers of material or materials. Some packaging containers now combine plastic with both paperboard and metal foil. A plastic food-packaging film might look as if it contains only one layer, but in reality it could contain six or more very thin layers of different plastics.

The purpose of developing composites and laminates is to combine the physical properties of the components within one material. The idea is not new. In the British Isles small hide-covered wicker boats originated in ancient times and are still used by fishermen. Plywood is a more modern example. Some multilayer food-packaging film has one layer to provide strength, another to provide a moisture barrier, and still others to provide an aroma barrier and a printing surface. Composites and laminates give added dimensions to materials science in the search for an ever-increasing range of properties and versatility of application.

The Distinction of Plastics as a Material

Plastics have successfully competed with other materials in so many markets partly on account of their low cost, especially in their processing. Two simple market examples are the gears found in toys and small devices, and zippers in clothing. Previously made of metal, they are nowadays often made of plastic at less cost. The plastic gears and zippers perform as well as metal in many applications, and the lesser durability of the plastic often does not matter because the toy or clothing is discarded—in our disposal-oriented society—long before the plastic gear or zipper fails.

Plastics also compete well in many markets because of their superior performance. For packaging applications, plastic is less breakable than glass and of lighter weight. Plastic can keep liquids cold longer than metal or glass. And it is easy to imagine how happy George Washington would have been to trade in his now famous wooden dentures for the vastly superior modern version, made of plastic.

The durability of plastics has also helped them in the marketplace, and maximizing the durability of plastics has been a major goal of plastics research.

Materials and the Ecosystem

Today another property of materials has taken on greater significance than in the past—the relationship between materials and the ecosystem. We now ask questions about the materials we use that we did not ask in the past. Where does the material come from? Where does it go after it is no longer in use? What environmental price is associated with its use?

Plastics are being given a careful look as a result of heightened concern over the environment, and some properties of plastics are being reevaluated in terms of how the large use of plastics affects the environment.

3

Plastics and the Environment

Raw Materials

With the increased use of plastics, people have become concerned over the impact plastics have on the environment. One source of concern is the raw materials from which they are produced. Virtually all plastics are made from petroleum (crude oil), natural gas, and coal. Those raw materials are the **feedstocks** of the plastics industry. They are natural resources that have taken millions of years to be formed, and are nonrenewable.

Plastics feedstocks account for only 4 or 5 percent of oil and gas production; processing energy uses another 2 or 3 percent. Some take these small percentages to mean that the use of fossil feedstocks for plastics represents no independent or additional drain on resources; some refer to plastics as the by-products, or even waste products, of fuels production. However, our dependence on fossil fuels for energy goes hand in hand with a dependence on them to provide the feedstocks for the production of plastics. Maintaining plastics markets, let alone increasing them, requires corresponding levels of fuels production, and economic pressures in one area cause pressures in the other; plastics production and world oil prices are often correlated. For countries that import oil and have large plastics industries, like the United States, part of the dependence on foreign oil—albeit a small part—has to be considered the result of plastics production.

No one knows how much oil, natural gas, and coal exist on the planet, but the known reserves and estimates of unknown reserves have been tallied. Coal is the most abundant of the **fossil fuels**, and there is possibly enough coal to supply the world's needs for two or three centuries at present rates of use. Petroleum is not so plentiful. Worldwide reserves have been estimated to be in the range of 200 billion tons, which is enough for only fifty years or so at the current rate of consumption. The supply of natural gas is sufficient for approximately the same length of time, barring new discoveries. Moreover, these resources are not evenly distributed throughout the planet; most oil reserves are in the Middle East. The United States's supply of oil may last as few as ten years.

The first major turning point, though, will not be when the world supply of oil is exhausted—it may never become totally exhausted— but when world production of oil peaks. The level of world stress that results from that occurrence will depend on how much thought and planning will have been given to the problem beforehand. But the use of our limited and nonrenewable supply of fossil resources for the large-scale manufacture of plastics is a legitimate environmental concern now, because it increases the rate at which we approach a day of reckoning some time in the future.

Plastics Waste

The second major environmental concern is plastics **waste**, both managed waste and litter. The concern over plastics waste is not new. In the 1960s it was suggested that so much plastic had been manufactured that the planet could be wrapped in it. It is not the use of plastics, but the *magnitude* of the use that has been the cause of concern.

In the United States over 60 billion pounds of plastic are discarded into the waste stream each year. (In 1970 the plastics waste stream was only 4 billion pounds.) In Western Europe over 35 billion pounds of plastics waste are generated each year. Well over half the plastics waste stream is in **municipal solid waste (MSW)**.

Municipal solid waste includes common garbage or trash generated by homes, businesses, institutions, and industries. Municipal solid waste accounts for only a few percentages of the total waste stream, but most people are aware of it more than other wastes because everyone generates it, its collection is highly visible, and people pay for its handling in a direct way. The United States generates over 400 billion pounds of municipal solid waste each year, amounting to more than 4 pounds per day per person, and accounting for almost 50 percent of the world's total.

Plastics now make up a significant part of a typical municipal solid-waste stream, and represent the fastest growing component. Forty-four billion pounds of plastics enter the United States's municipal solid-waste stream each year, equivalent to half a pound per day per person. In the United States, plastics on average account for 10 percent of the total weight of municipal solid waste—more than metals (8 percent) and more than glass (6 percent). In Europe municipal solid waste is about 5 percent plastics by weight, and in Japan 17 percent.

Plastics account for around 18 percent of the volume of municipal solid waste. Excluding **organic** waste—paper, yard waste, food waste, and wood—plastics make up 30 percent of the remaining weight and as much as one-half the remaining volume.

Worldwide about one-half of all discarded plastic comes from packaging. Almost one-third comes from packaging that is discarded soon after use, typically in less than a year, and sometimes much sooner. Heavy-duty plastic construction materials have a longer life span, which means that much of the plastic that was put into use decades ago, when plastics use began in earnest, will start showing up in increasing amounts in the near future.

Plastics waste is generated on the sea as well as on land; millions of pounds of plastics waste are produced at sea each year. Government regulations now ban the dumping of plastics at sea, but adhering to the regulations is expensive.

As litter, plastics also present problems. Plastic litter is hazardous to a variety of living creatures; birds, fish, and other animals die from ingesting it or becoming entangled in it. It is also unsightly and disturbs our enjoyment of nature. Litter constantly accumulates on highways. Beach litter is between 40 and 60 percent plastic. Ocean beach litter is often not even of local origin, but floats in from the sea. A major cleanup along the Texas coast once produced thousands of plastic bags.

Gathering litter so as to have it enter the waste stream of managed programs is always expensive, even for relatively accessible litter, and for remote litter it can be prohibitively expensive. The cost of gathering litter makes it likely that there is always going to be a litter problem to some degree or other.

Although the costs of waste management used not to be highly visible and used not to be a matter of great public concern, that is no longer true. Waste management has become a large and expensive affair—New York City taxpayers spend over a billion dollars each year for waste management. Because of the expense, the general public, worldwide, is growing increasingly sensitive to waste-management problems, and public concern has periodically erupted in the form of local restrictive legislation. Plastic dishes for school lunches were once banned in Tokyo, resulting in a change to stainless steel. Plastic garbage bags have at times been banned by local governments. Whatever the frequency, duration, or impact of these local restrictions, they dramatically illustrate a public concern that is both significant and serious. But what place do plastics have in the overall waste-management picture?

Managing Plastics Waste

The early stages of managing any kind of waste are embodied in the general motto of environmental conservation, "Reduce–Reuse–Recycle." There are limits to what can be expected of plastics waste management in all three categories. The term **source reduction** refers to the reduction of the amount of materials entering the waste stream by redesigning patterns of production or consumption. Consumer restraint probably cannot be counted on for major source reduction as long as plastics are so inexpensive and versatile. Forced reduction through restrictive legislation is not something people want to rush into.

Some proposals to replace plastic packaging with other materials—like metals, glass, or paper—become environmentally less attractive after they are given a closer look. Metals are not a renewable natural resource, and they are much more expensive than plastics. Metals are durable and can be reused and recycled, but in many applications the number of reuses and recyclings would have to be unrealistically large for metals to be economically competitive with plastics.

Glass is made from sand, which is abundant and cheap, and glass is reuseable and recyclable. On the other hand, the energy requirement for processing sand into glass is much higher than for the production of plastics. Once glass enters the waste stream, it presents its own problems. Glass, like plastic, eventually ends up occupying landfill space. If it ends up as litter, glass is as unsightly as plastic and may well be as hazardous.

Paper and cardboard contribute less to the waste-management burden, but they are made from wood, which is not a rapidly renewable resource—accelerated deforestation would not be an environmental improvement. And producing paper from wood is an energy-intensive process. So a simple shift away from plastics to other present packaging materials cannot be counted on to eliminate the environmental impact of using plastic packaging.

I do not intend to underestimate the complexities of environmental impact analysis but, for example, even Martin B. Hockings's detailed and often-cited impact analysis comparing a hot-drink paper cup with a polystyrene-foam plastic cup suggested that the environmental impact of the paper cup is not overwhelmingly less than that of the plastic cup when both resource use and energy use are included.

Excessive packaging has at times come under fire from the general public, forcing manufacturers to rethink and simplify packaging design. Such pressures may contribute to limiting extravagant packaging applications in the future. Moreover, companies have always been motivated to reduce and simplify packaging as a way of reducing packaging costs. Through improved design the plastic milk jug of twenty years ago has been replaced by today's plastic jug, which weighs 45 percent less. Plastic grocery bags use 70 percent less plastic than they did in 1976. But even with continued design improvements, packaging needs will remain very, very large.

The **reuse** strategy also has limitations. Many plastics applications are not amenable to reuse because of impurities introduced in the original use. Agricultural covers and waste bags are two examples. Food packaging is another. Disposable diapers are not reusable; they account for millions of pounds of plastics use each year. In all these cases the plastics will enter the waste stream quickly.

Recycling has come to be important for plastics. Plastics producers and processors have long been recycling internal scraps generated during production. Some scrap allows immediate reuse in the original process; other scrap requires further processing and different end-use applications, but the recycling is efficient on account of the ease of collection and separation by type. In-plant, or preconsumer, recycling amounts to several billion pounds a year.

Wide-scale postconsumer recycling of plastics is relatively new, but in many geographical areas it is now well established. There are both curbside pickup programs and drop-off recycling stations for a wide variety of plastics. Some manufacturers have postconsumer recycling programs for their plastic products and for the plastic packaging they use in distributing goods. The large company 3M offers to recycle used transparencies returned to it. It estimates that 15 million pounds of used transparencies are dumped into landfills each year.

The recycling of plastics has limitations, however. Collection and sorting are only the first steps of plastics recycling, and the technology of recycling is still being developed, especially the use of compatibilizers and stabilizers in the recycling of mixed plastics. Composites and laminates complicate recycling efforts. As plastic compositions become more complex the recyclability of some plastics may even decrease unless recycling technology keeps apace. On the other hand, simple plastic formulations have a higher chance of being recycled, and packaging designers are now being encouraged to keep packaging compositions as simple as is compatible with maintaining adequate properties.

Some plastics are far more difficult to recycle than others no matter what their application. These include the thermosetting plastics, which cannot be softened and reshaped through heating. They amount to about 10 percent of plastics production.

Deterioration of the material occurs during recycling; the polymer chains in the plastic are sheared, and the impact strength is often reduced. Polystyrene, after three cycles through an extruder machine, suffers a 9 percent reduction in the size of its polymer chains and a 34 percent reduction in impact strength.

Recycled plastics therefore have to be used in less demanding applications than the original product made of virgin material. (An exception is when recycling involves recovery of the original raw materials from which the plastic was made, which allows regeneration of virgin plastic.) Examples of uses for recycled plastics are plastic lumber, outdoor furniture, pipe, hose, toys, weather stripping, car trunk liners, ski jacket insulation, carpeting, and egg cartons, where performance demands are not as great as in many other applications. Products made from recycled plastics are themselves less likely to be recycled, so that recycling represents a "spiraling down" to some other disposal method.

Mass loss during processing also limits recycling. With a processing efficiency of 90 percent, it would take only three recyclings to reduce the material to 73 percent of its initial amount.

Recycling efforts suffer from unfavorable economic factors as well, including transport logistics, energy costs, and unsteady markets. Of all the components of a recycling program, developing markets for recycled materials has proved to be the most difficult to manage.

In spite of these problems, and at significant cost, postconsumer plastics recycling now has a firm foothold in waste management. Collection programs for recycling plastics now exist in 59 percent of the communities in the United States and are available to 80 percent of the population. The number of curbside programs has grown from slightly more than 1,000 in 1988 to over 9,000 in 1998; they are available to an estimated 140 million people. The number of drop-off sites in the United States has increased to nearly 13,000.

In the United States over 2 billion pounds of postconsumer plastics, principally beverage containers, are separated from the municipal solid-waste stream and recycled each year. In 1996 more than 600 million pounds of the recycled plastic was poly(ethylene terephthalate), used in soft-drink bottles. In Western Europe over 2 billion pounds of postconsumer plastics are also recycled. Plastics recycling from other waste streams totals approximately the same amounts.

On the other hand, the several billion pounds of plastics being separated from the waste stream and recycled each year in the United States and Western Europe, although substantial, are nevertheless a small percentage of the tens of billions of pounds of plastics waste generated. Over 90 percent of plastics waste is not recycled and moves on to later-stage disposal methods.

Incineration is sometimes referred to as *energy recycling* because of the possibility of generating heat. But it destroys material value and does not reduce dependence on virgin raw materials. If considered to be recycling, incineration might also be called last-resort recycling. Incineration accounts for the disposal of about 17 percent of municipal solid waste in the United States, but there is large variation by region—from 1 percent in the western states to 45 percent in New England. In Western Europe an average of about 19 percent is incinerated, but there the amount also varies among countries—from about 5 percent in Great Britain to 80 percent in Denmark. In Japan, land is so scarce that landfilling is practically impossible, and over 90 percent of the solid-waste stream is incinerated.

Incinerators may be able to function cleanly with the best technology, and the energy generated has value if it is harnessed. But incineration is not always acceptable to a majority of voting residents in a community. Their concern comes from the toxicity of the potential pollutants, including hydrogen chloride, heavy metals, and dioxins. Technology exists for dealing safely with such chemicals, but managers have to have the substantial funds needed to provide, and maintain, adequate equipment and to adhere continually to the required operating practices.

New technologies are emerging, based on **pyrolysis**. In pyrolysis, waste material is heated in the absence of oxygen, and possibly under pressure as well. The process drives off volatile components, including gaseous hydrocarbons, liquid hydrocarbons, and a mixture of other chemicals whose composition depends on the nature of the original waste. Usable fuels and other chemicals are generated as well as heat. Pyrolysis, unlike incineration, preserves some of the material value of the waste.

One of the newest pyrolysis technologies is directed at *cracking* waste plastics, a process in which the long polymer chains composing the plastics are broken up into smaller chains. A polymer cracking plant would convert the waste plastics into the hydrocarbon building blocks from which the plastics were originally made. Those hydrocarbons could then be reused as polymer feedstocks, making pyrolysis a form of *feedstock recycling*.

In one approach, plastics waste is pretreated, including size reduction and metal removal. It is then fed into a cracking reactor where the plastic components are thermally broken down into a mixture of lower molecular weight hydrocarbons, which comes off the reactor as a hot gas stream. The hot gas stream is brought into contact with a circulating liquid, producing a liquid distillate—representing 80 percent of the plastics waste stream, an uncondensed fraction of light hydrocarbon gas (of 10 to 15 percent) and a remainder of ash and coke. Some of the gaseous fraction is recycled as fuel for the reactor.

The distillate is then used as a feedstock in cracking units of the type already in standard use within the petroleum refinery and petrochemical industries. The final products typically include about 30 percent liquified petroleum gas (LPG)—a mixture of low boiling point hydrocarbons that become liquid under pressure. That fraction includes the low molecular weight compounds (olefins) that can be used to manufacture virgin polymers. Other final products include a gasoline fraction, amounting to about 50 percent, and smaller amounts of diesel fuel, fuel oil, fuel gas, and coke.

The technology has not yet been fully developed, but interest in the technology is growing and its feasibility is being tested in pilot plant studies. One of the difficulties is the poor thermal conductivity of plastics; heat introduced to the cracking reactor does not easily flow through the material. The time needed to raise and maintain the high temperatures required—around 500 °C—in a controlled manner is increased and there is a risk of fouling problems. The cracking reactor has to be chosen to maximize heat transfer, which constitutes a complex engineering design problem.

If the technology proves successful it would represent genuine feedstock recycling, producing feedstocks for virgin plastics. The technology is significant because it avoids the destruction of material value that characterizes incineration.

Landfills are the last stop for the waste stream. They represent thoroughly devalued resources and they pose risks, including the risk of toxins leaching into the water supply, the risk of long-term subsidence, affecting later use of the land, and the risk of uncontrolled production of methane.

Methane that is retrieved from vented landfills has value as a fuel, but the anaerobic decomposition of garbage in landfills is limited and variable; some have argued that landfills are better described as mummifiers than as composters. Landfills are vented partly to prevent the future accidental ignition of methane in populated areas built over or near old landfills. In 1981 several small explosions near a landfill in Port Washington, Nassau County, New York, triggered state and federal studies of the problem. All New York State landfills built after 1988 have been required to have collection systems for methane. Accidental ignition apparently also occurred at a 1986 pop concert in Mountain View, California, when a member of the audience, attempting to light a cigarette, sparked a five-foot column of flame. There were no serious injuries, but a woman sitting nearby, whose hair was singed, filed a lawsuit.

Apart from whatever other problems plastics contribute to landfill management, they take up a lot of space, even when compacted. Landfills will fill up partly on account of plastics. As landfills become full, the problems of finding new sites, getting site approval, developing the site for landfill use, and establishing disposal fees have become very time consuming, controversial, and expensive. In some European countries, including Germany and the Netherlands, the landfilling of plastics is no longer even allowed.

Between 1988 and 1998 the number of landfills in the United States dropped from 8,000 to 2,314—but the average landfill size increased because it is often the smaller landfills that are closed. In New York State there were 550 landfills in the mid-1980s, but many of the smaller landfills have been closed without replacement, leaving around 30. Now that New York City has closed Fresh Kills, its primary landfill on Staten Island, it must cope with several large problems. Not the least of these is the need to export garbage to more expensive disposal facilities, at an expected increase in export disposal costs from $31 million in 1998 to $240 million in 2002, bringing the city's total waste-management cost to more than a billion dollars a year.

It is not that there is no land for landfills per se. One analyst estimates that all of the garbage that will be produced in the United States for the next thousand years would fit into one-tenth of 1 percent of the land area. The problems in siting new landfills have to do with minimizing transport costs, overcoming the objections of nearby residents, and satisfying a vast array of environmental regulations requirements.

Plastics are not responsible for the entire waste-management problem, but they are part of it.

Environment Friendly Plastics?

Can plastics be made from feedstocks other than fossil feedstocks? Can plastics be made that pose less of a waste-management burden than current plastics? The thesis of this book is that *environment friendly plastics*—green plastics, if you will—is not a contradiction in terms.

With respect to fossil resources, efforts have been under way for some time to conserve fossil fuels by increasing the use of alternative energy sources, including hydroelectric, solar, geothermal, and wind sources, and—in some parts of the world—nuclear energy. Renewable planetary **biomass** is also regarded as a potentially large energy source. Biomass refers to the total mass of matter generated by the growth of living organisms, including plants, animals, and microorganisms (but animals make up only 1 percent of the total biomass). Estimates of the amount of new biomass produced each year range from 100 to 200 billion tons. Approximately two-thirds of the total is from the land, with the remaining third coming from the water.

The large-scale use of ethanol for energy is being evaluated, for example, because it can be produced from annually renewable agricultural products such as corn syrup or cane sugar. Conversion of renewable biomass to fuel presents the possibility of supplying an unending stream of fuel long after fossil fuels have been exhausted.

Are there alternative, and perhaps new, materials for some of our uses of present plastics that do not drain our limited—and therefore valuable—supply of fossil resources? Can we look to the renewable planetary biomass as a potential feedstock for the production of plastics, just as we are now looking to it as a partial alternative to fossil fuels for energy production? Is it possible to slow down our use of fossil fuels to produce plastics, so as to make those natural resources last longer? And, most important, since fossil fuels will inevitably become more and more expensive as the world supply decreases, is now not the time to begin developing technologies based on alternative feedstocks? We cannot expect to develop adequate technologies overnight. Questions like those have led to a search for new kinds of plastic made from *renewable* feedstocks.

With respect to managing plastics waste, one basis for public concern is that current plastics are **recalcitrant**, or resistant to degradation. As litter they add to the hazards and nuisance of other nondegradable litter—like glass and metals. As managed waste their inertness at first made it seem that they were headed for incineration or landfill disposal immediately after use ("burn and bury") until recycling came to be a real possibility. But even now, plastics that are not recycled end up in incinerators or landfills.

Are there alternative, and perhaps new, materials for some of our uses of plastics that, after reuse and recycling had run their course, would be more amenable to environment friendly late-stage waste-management strategies?

For example, if new plastics were formulated and processed to be clearly incineration safe, then incineration and pyrolysis, with their speed and energy efficiency, could become more easily acceptable and more widely adopted forms of disposal.

If plastics derived from biomass were developed, **composting** could provide another avenue for their disposal. Composting is a *managed* process that controls the microbial decomposition of biodegradable material, and its transformation into carbon dioxide, water, minerals, and relatively **stabilized** humus-like organic matter called compost. (The word *stabilized* refers to the point at which microbial activity reaches a low and relatively constant level.) **Humus**, the product of the *natural* (unmanaged) process, is the dark organic portion of soil remaining after the partial decay of plant and animal matter brought about by prolonged microbial decomposition. In composting, the natural biological process is used to stabilize mixed organic material recovered from waste. Composting returns the fundamental components of waste to their natural **biogeochemical cycles**. It is sometimes referred to as "natural" or "biological" recycling, or as a fourth R in the motto of environmental conservation, "Reduce–Reuse–Recycle–Return to Nature."

Compost has value as a soil amendment for use in agriculture, horticulture, sod production, and landscape architecture. Markets for compost include farms, greenhouses, plant nurseries, garden centers, orchards, golf courses, soil blenders, municipalities, and the general public.

Composting technologies are developing rapidly, and composting times have been steadily decreasing until times of only several months are now a realistic expectation. Methods of composting include mechanical mixing and aerating, dropping materials through a vertical series of aerated chambers, or placing the compost in piles—windrows—out in the open air and mixing it or turning it periodically. Composting could vastly reduce the use of incinerators and landfills, but current plastics, with rare exceptions, are not compostable (Chapter 5).

Degradable plastics would have special value as *bagging* for compostable garbage, including yard waste and food waste. Noncompostable bags have to be emptied, either by hand or by machine, so as not to contaminate the compost. But both methods of debagging add to the cost of the compost, and disposal of the emptied bags further adds to the cost. If the debagging is not 100 percent efficient, the value of the compost is reduced. Compostable bags would eliminate such problems.

New composting technologies are under development in which waste biomass—rather than simply being converted to compost for soil enrichment—is converted, with the addition of microorganisms, into chemicals, including alcohol fuels. The alcohols could be burned in place of gasoline.

Anaerobic composting is the result of bacteria that thrive on an oxygen-free environment at nearly 95 °F. It currently accounts for only 5 percent of commercial composting, but this percentage may grow as new technologies are developed. One current anaerobic composting project aims to test the feasibility of composting as much as 39,000 tons of yard-trimming feedstocks a year at one facility. Yard trimmings are mixed with water in a three-to-one ratio, and after about sixty days produce a mix of liquid fertilizer, nitrogen-rich compost, and methane.

Composting has value well beyond that of managing waste. There is growing realization that in the future its most important value will be in supporting long-term sustainable agriculture. Crop land is steadily being lost to development, and topsoil is continually lost to erosion. We have been depending on continued increases in agricultural productivity, and composting can help.

Compost increases organic matter in the soil, feeding soil microorganisms and increasing water- and nutrient-holding capacities; it provides a slow release of nutrients; it suppresses plant pathogens and provides an alternative to raw manures that may introduce pathogens; and it reduces erosion. The end results are increased crop yield and quality, reduced dependence on chemical fertilizers and pesticides, and reduced need for water.

Composting infrastructures have been developing rapidly in the United States and in some European countries, notably Germany. The United States has more than 3,000 modern composting plants; Germany has over 400. In Germany, 15 billion pounds of waste are composted annually.

In the United States, there is growing interest in the collection and composting of organic waste other than yard waste—and for good reason. The municipal solid-waste stream is typically 38 percent paper and paperboard, 13 percent yard trimmings, 10 percent food residues, and 5 percent wood, so that over two-thirds of the weight of the municipal solid-waste stream, not otherwise recyclable, could be composted using existing technologies.

Source separation is also proving feasible. In Europe some individual regions have achieved separate collection of up to 30 percent of the organic fraction of solid waste. Helsinki, Finland, has achieved total solid-waste source separation, and the organic fraction is totally composted.

Residential curbside collection of organic compostables is virtually nonexistent in the United States except for yard trimmings and wood. But many states—interested in increasing their overall recycling rates—have begun developing composting infrastructures for commercial and institutional sources, like grocery stores, restaurants and cafeterias, hotels, hospitals, schools and universities, prisons, food processors, and others. Pilot projects exist in California, Massachusetts, New York, North Carolina, Oregon, Texas, Vermont, Washington, and other states. The New York correctional system uses composting at thirty prisons across the state to divert 12,000 tons of food residuals a year. Vermont is testing the feasibility of the curbside collection and composting of household food residuals and nonrecyclable paper. Through such programs a nationwide composting infrastructure is developing

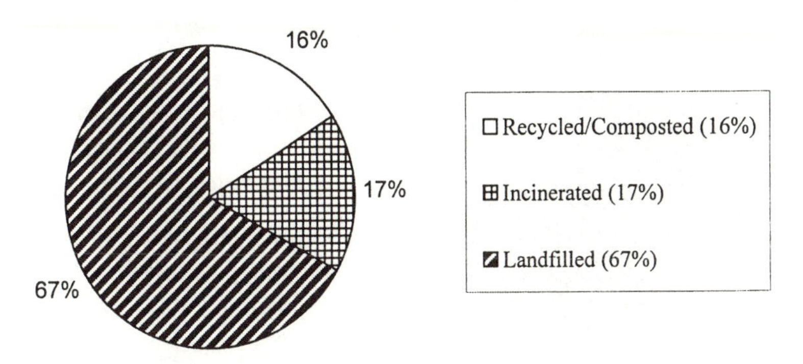

Figure 3.1 *Disposal of municipal solid waste in the United States in 1990*

and feedstock availability is expanding and diversifying. That growing composting infrastructure would be available for the disposal of degradable plastics.

There are indications that the public would be receptive to degradable plastics. The seriousness with which waste management is now viewed by the public is reflected in the dramatic change in how municipal solid waste is disposed of in the United States. In 1970 most waste was simply sent to open dumps, 10 percent was disposed of in sanitary landfills, about 10 percent was incinerated, and 7 percent was recovered for recycling. By 1990, when the municipal solid-waste stream had grown by 70 percent, open dumps were no longer legal and most of the municipal solid-waste stream was landfilled (fig. 3.1).

By 1998 the profile had continued to change, partly on account of great public interest (fig. 3.2). The percentage of municipal solid waste being landfilled had fallen from 67 percent to 55 percent. The amount being incinerated had remained virtually unchanged, at around 17 percent. The amount being recycled, including yard trimmings composting, had increased from 16 percent to 28 percent. In the same period the number of curbside recycling programs had increased more than eightfold—to 9,300—and the number of yard trimmings programs had risen more than fivefold—to 3,800. Few people had believed such a change was possible by the turn of the century.

We keep coming back to the same question. Are there plastics that can satisfy our needs, and desires, for at least some of the

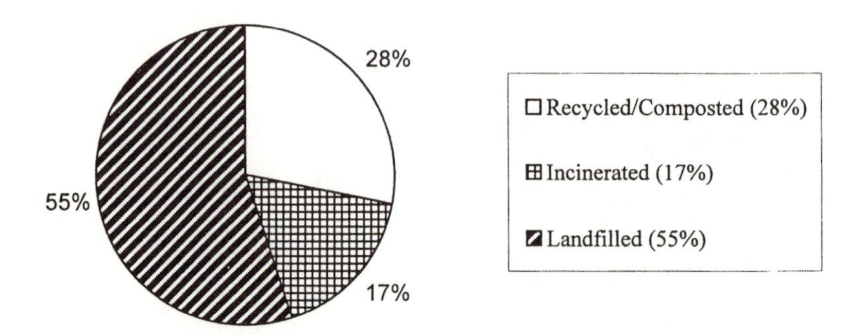

Figure 3.2 *Disposal of municipal solid waste in the United States in 1998*

products now available, but also satisfy our legitimate environmental concerns? A partial practical solution to these environmental concerns might be to use materials that are derived from rapidly renewable resources and that are degradable in an environmentally benign manner by composting or safe incineration.

Some argue that developing plastics that degrade in an environmentally benign manner runs counter to the environmental goal of recycling, on the grounds that the very existence of degradable plastics provides a disincentive for recycling efforts. But that argument ignores the large amounts of plastics now used in applications where no significant reuse or recycling can realistically be expected to take place. And it ignores plastics that have very limited recyclability and are destined for early incineration or landfilling. In those applications, seeking a degradable alternative is the environmentally sound approach. Others are afraid that a litterer might feel greater license to litter if plastics are degradable. That fear, however, attributes a greater complexity to the mind of the litterer than is actually there. A litterer simply litters.

Before considering what the composition of an environmentally degradable plastic made from renewable resources might be, the chemical nature and environmental degradability of current plastics have to be examined, so that the differences between the two can in the end be fully appreciated.

4

The Chemical Nature of Plastics

Polymers

Essential ingredients of plastics are polymers. A **polymer** is a substance consisting of molecules characterized by the repetition of one or more types of *monomeric unit*. An **oligomer** is a substance composed of only a few monomeric units repetitively linked to each other, such as a dimer, trimer, or tetramer or their mixtures. (The word polymer comes from the Greek *polys*, meaning many, and *meros*, meaning part. Similarly, **monomer** is from the Greek *monos*, meaning single or sole, and oligomer is from *oligos*, meaning few or small.)

To maintain the distinction between the polymer substance and the polymer molecule, of which the substance is made, the word molecule is typically included explicitly when needed; for example, polyethylene, the substance, as distinct from the polyethylene molecule.

The size of a polymer molecule is measured in terms of **molecular weight**, which is the sum of the **atomic weights** of all the atoms in a molecule. Atomic weights are measured on a scale in which one unit weighs 1.66×10^{-24} grams. In these units the molecular weight of a polymer might be as large as a million or more.

Polymers differ chemically from one another in the types of atom they contain, in the number of atoms, and also in the arrangement of the atoms within the polymer molecule. The most common atoms in polymers, and their chemical symbols, are carbon (C), hydrogen (H), and oxygen (O). Some polymers contain chlorine (Cl), fluorine (F), nitrogen (N), or sulfur (S) atoms. Other atoms are sometimes also present, but in smaller numbers. Different polymers can be distinguished by their *chemical formulas*.

A *chemical formula* expresses the relative number of each kind of atom that is present in a chemical compound. The formula for water, H_2O, indicates that a water molecule contains two hydrogen atoms and one oxygen atom. A more elaborate type of chemical formula, a *structural formula*, shows not only what atoms a molecule contains, but also how the chemical bonds connecting them are arranged. A structural formula for water, for example, is H–O–H, indicating that there is a central oxygen atom bonded on either side to a hydrogen atom. The line connecting two atoms represents a pair of bonding electrons.

A structural formula might show additional details of the geometric arrangement of the atoms, including bond angles. The structural formula for water shown in (a) in figure 4.1 indicates that the three atoms in a water molecule are not arranged in a straight line.

Bonds that contain four electrons instead of two, and are particularly strong, are represented with two lines connecting the atoms instead of one. Formaldehyde (CH_2O), for example, shown in (b) in figure 4.1, contains a carbon-oxygen double bond. In methane (CH_4), the carbon-hydrogen bonds are arranged in a tetrahedral geometry about the central carbon atom, as shown in (c) in figure 4.1, but its structural formula is often drawn as a projection onto a plane for simplicity, as shown in (d) in figure 4.1.

A common symbol for the benzene molecule (C_6H_6) is shown in (e) in figure 4.1. The molecule is made up of a ring of six carbon atoms, each bonded to a hydrogen atom. The corners of the hexagon represent carbon atoms; for simplicity the hydrogen atoms are not shown. The circle inside the ring indicates the *aromatic* nature of the benzene ring; the carbon-carbon bonds are intermediate in strength and length between single and double carbon-carbon bonds.

More detailed drawings might indicate the relative sizes of the atoms, and sometimes the three-dimensional shape of the molecule can be hinted at using perspective techniques.

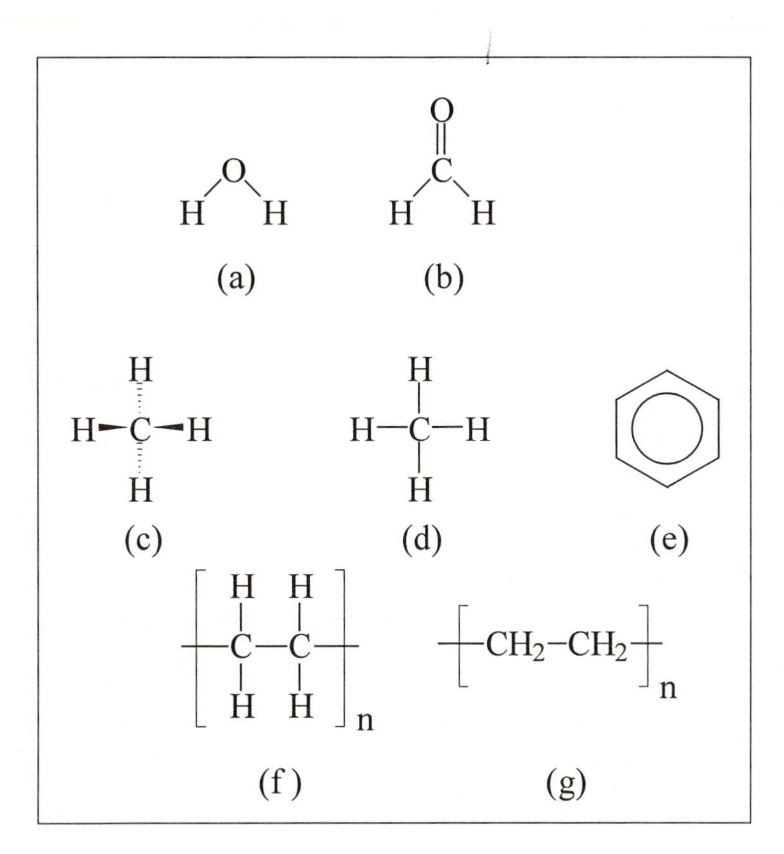

Figure 4.1 *Structural formulas of (a) water, (b) formaldehyde, (c) and (d) methane, (e) benzene, and (f) and (g) high-density polyethylene*

In chemistry, **organic** compounds are generally those that contain carbon atoms combined with other elements. Inorganic compounds do not contain carbon—although some simple compounds containing carbon, like carbon monoxide, carbon dioxide, carbonates, and a few others, are generally considered to be inorganic; they contain "inorganic carbon."

Continued on next page

Continued from previous page

Organic compounds are classified according to their chemical composition. *Hydrocarbons* contain only carbon and hydrogen atoms. They are subdivided according to the type of chemical bonds in the compound. *Alkanes* are hydrocarbons in which each carbon atom is bonded to four other atoms with single bonds; they are thereby said to be *saturated.* Methane, shown in (c) and (d) in figure 4.1, is the simplest alkane. *Alkenes,* sometimes called *olefins,* are hydrocarbons that contain at least one carbon-carbon double bond. *Alkynes* are hydrocarbons that contain at least one carbon-carbon triple bond. *Aromatic* hydrocarbons are characterized by the presence of a benzene ring, shown in (e) in figure 4.1, or a related structure. Alkenes, alkynes, and aromatic hydrocarbons are *unsaturated.*

Other organic compounds can be considered *derivatives* of hydrocarbons, in which one or more hydrogen atoms have been replaced by one or another *functional group,* composed of an atom or group of atoms having a characteristic chemical reactivity. If a hydroxyl group (–OH) replaces the hydrogen atom, the compound is an *alcohol.* The formula of an alcohol can be expressed simply as ROH where R stands for the hydrocarbon—minus the replaced hydrogen atom. Alcohols are named on the basis of the parent hydrocarbon, as in methyl alcohol (CH_3OH) or ethyl alcohol (CH_3CH_2OH). For all but the simplest compounds a *condensed structural formula* can be used, as for ethyl alcohol (C_2H_5OH).

In alcohols, if the carbon atom bearing the –OH group is itself bonded to only one carbon atom, the alcohol is a *primary* alcohol as in propyl alcohol ($CH_3CH_2CH_2OH$). In *secondary* alcohols the carbon atom bearing the –OH group is bonded to two carbon atoms, as in isopropyl alcohol, $(CH_3)_2CHOH$. (A notation extension is introduced here in which it is understood that the central carbon atom is bonded directly to all four groups; i.e., the molecule contains two $C–CH_3$ bonds, a C–H bond, and a C–OH bond.) In *tertiary* alcohols, correspondingly, the carbon atom bearing the hydroxyl group is bonded to three carbon atoms, as in tertiary butyl alcohol, $(CH_3)_3COH$.

Ethers have the general formula R–O–R' where R and R' are the two parent hydrocarbons (each minus a hydrogen atom), and R and R' may be the same, as in methyl methyl ether, or different, as in

Continued on next page

Continued from previous page

methyl ethyl ether. An *epoxide* is a cyclic ether containing a three-membered, oxygen-containing ring. The simplest epoxide is ethylene oxide,

$$\overset{O}{\diagup\ \diagdown}$$
$$CH_2-CH_2$$

Amines can be considered derivatives of ammonia, NH_3, in which one or more of the hydrogen atoms are replaced with –R groups, as in primary (RNH_2), secondary (RR'NH), or tertiary (RR'R"N) amines.

Several classes of organic compound contain the *carbonyl* group, a carbon atom doubly bonded to an oxygen atom,

$$>C=O$$

In *aldehydes* the carbon atom is bonded either to two hydrogen atoms as in formaldehyde (CH_2O), shown in (b) in figure 4.1, or to one hydrogen atom and an –R group, as in acetaldehyde (CH_3CHO). In carbonyl compounds the carbonyl carbon atom and the three atoms bonded to it lie in the same plane.

In *ketones,* RCOR', the carbonyl group is bonded to two –R groups, R and R'. R and R' may be the same, as in methyl methyl ketone, (CH_3COCH_3; i.e., acetone), or different, as in methyl ethyl ketone, ($CH_3COCH_2CH_3$). Replacing the –R' group of a ketone with an –OH group yields a *carboxylic acid,* denoted in abbreviated fashion as RCO_2H. Replacing the –R' group of a ketone with an –OR' group results in an *ester,* RCO_2R'. In *amides,* the carbonyl is attached to an –R group and an amino group (–NH_2, –NHR, or –NRR'). In *imides,* two carbonyl groups flank an NH group (–CO–NH–CO–).

A molecule may be *polyfunctional,* containing two or more reactive groups. The groups may be of the same kind as in glycerol, $CH_2OH–CHOH–CH_2OH$, or they may be different, as in glycolic acid, $CH_2OH–COOH$, or the amino acid, glycine, $NH_2–CH_2–COOH$.

Although a polymer molecule has a large number of atoms, its chemical formula can often be represented very simply because the molecule consists of a single structural unit being repeated again and again. High-density polyethylene, for example, has the simple-looking structural formula shown in (f) in figure 4.1 or, simpler still, that in (g) in figure 4.1. The n means that there can be a large number (n) of the $-CH_2-CH_2-$ groups joined together to form a long polymer molecule. For simplicity, the chemical groups at the ends of the molecule are usually not included.

A **polymerization** is a chemical reaction in which low-molecular-weight monomer molecules react to form polymers. The number of monomeric units linked is the *degree of polymerization*. A polymer resulting from polymerization involving a single type of monomer is a **homopolymer**. A polymer formed from more than one type of monomer is a **copolymer**.

In **condensation polymerization** the chemical linking of the monomers is accompanied by the splitting off of a water molecule or other small molecule. The monomeric repeat unit in the polymer then lacks certain atoms that are present in the original monomer. An example is the polymerization of glycolic acid to produce the biodegradable polymer poly(glycolic acid).

$$n\ \underset{\displaystyle CH_2}{\overset{OH}{|}}\!-\!\overset{\displaystyle O}{\overset{\|}{C}}\!-\!OH \longrightarrow \left[CH_2-\overset{\displaystyle O}{\overset{\|}{C}}-O \right]_n + H_2O$$

In **addition polymerization** there is no such splitting off; the reaction often involves monomers containing carbon-carbon double bonds, as when olefins polymerize producing polyolefins. An example is the addition polymerization of ethylene to produce high-density polyethylene.

$$n\ H_2C{=}CH_2 \longrightarrow \left[CH_2-CH_2 \right]_n$$

The condensation-addition classification scheme for polymerization reactions is simplified. Polymer chemists use a more technical classification scheme to describe the complexities of polymerization reactions.

Some commercial polymers are not produced by direct polymerization, but by the chemical modification of another polymer. Poly(vinyl alcohol), for example, is produced from poly(vinyl acetate).

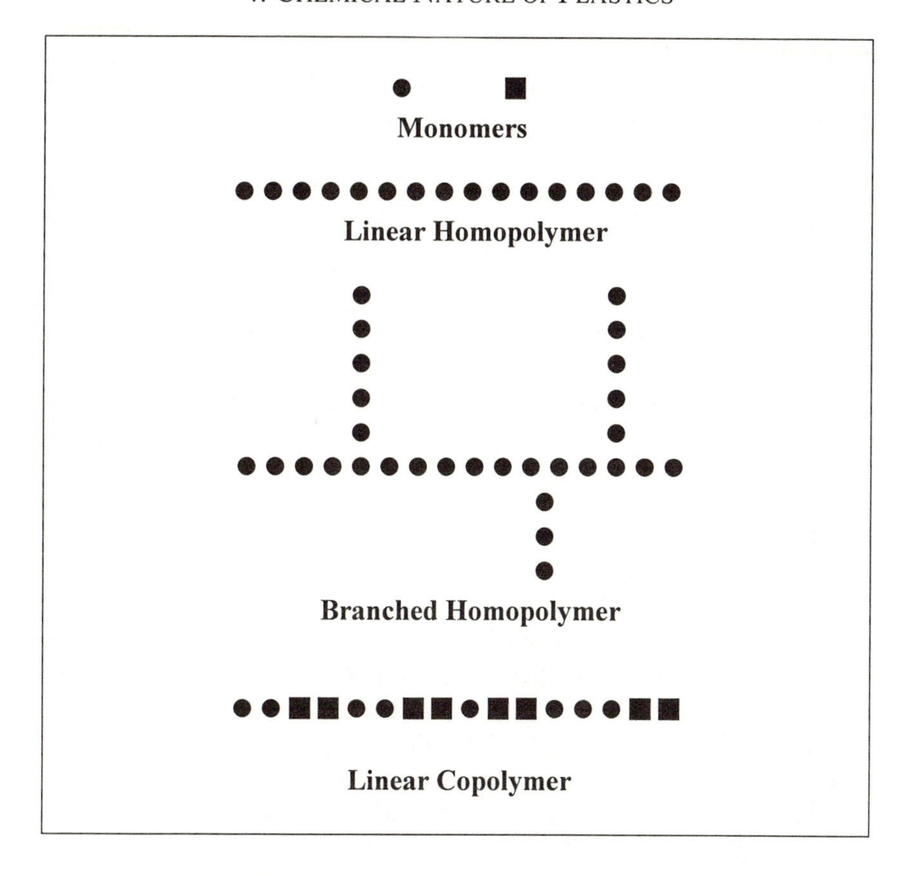

Figure 4.2 *Common types of polymer structure*

Because of the nature of polymerization reactions, the polymer molecules resulting from any particular polymerization will have a range of molecular weights, distributed about some average molecular weight. The molecular weight distribution plays a role in determining the properties of the polymer.

The repetitive nature of the chemical structure of polymers leads to polymer molecules being described as chain-like. Polymer chains can be either linear or branched, as illustrated in figure 4.2.

If the name of the monomer is written as a single word, the polymer is named simply by adding the prefix *poly*, as in polyethylene. If the name of the monomer contains more than one word, the name of the polymer is expressed by enclosing the monomer name in parentheses, as in poly(vinyl acetate). An example of a copolymer is poly(ethylene-*co*-styrene). Alternative conventions exist; in one, parentheses are used to enclose monomer names in all cases, and sometimes in common usage parentheses are omitted entirely.

Groups or classes of polymers are also given names, such as *polyolefins,* a monomer-based term which refers to the addition polymers of olefins (alkenes). *Poly(amino acids)* are the condensation polymers of amino acids. Names may also refer to the *repeating structural unit* in the polymer rather than the originating monomer. Polymers of amino acids, for example, are *polyamides,* and the condensation polymers of an acid and an alcohol are *polyesters.*

Plastics

A **plastic** is a material that contains as an essential ingredient one or more organic polymeric substances of large molecular weight, is solid in its finished state, and at some stage in its manufacture or processing into finished articles can be shaped by flow. Rubber, textiles, adhesives, and paint, which may in some cases meet this definition, are not considered plastics.

The starting bulk polymeric material, called **resin**, is processed so as to be formed into three-dimensional shapes, or into sheets or films. During the processing the resin often exhibits a degree of pliability, or plasticity, hence the name plastic. (The word *plastic* comes from the Greek *plastikos*, meaning able to be shaped.) After processing the material is often not very plastic, and may be quite hard. Various types of engineering processes are used, including

extrusion, a process in which heated or unheated plastic is forced through a shaping orifice (a die) in one continuously formed shape, as in film, sheet, rod, or tubing;

injection molding, the process of forming a material by forcing it, in a fluid state and under pressure, into the cavity of a closed mold;

compression molding, the method of molding a material already in a confined cavity by applying pressure and usually heat;

blow molding, a method of fabrication in which a heated plastic mass is forced into the shape of a mold cavity by internal gas pressure;

transfer molding, a method of forming articles by fusing a plastic material in a chamber and then forcing essentially the whole mass into a hot mold where it solidifies;

vacuum forming, a forming process in which a heated plastic sheet is drawn against the mold surface by evacuating the air between it and the mold.

A **cast film** is a film made by depositing a layer of plastic, either molten, in solution, or in a dispersion, onto a surface, solidifying, and removing the film from the surface. (Films can also be made by extrusion [see above] through a narrow die to form an *extrusion cast film*.)

Plastics are of two types. A **thermoplastic** is a plastic that repeatedly can be softened by heating and hardened by cooling, and in the softened state can be shaped by flow into articles by molding or extrusion. In thermoplastics the polymer chains are generally linear or only slightly branched. About 90 percent of the plastics produced today are thermoplastics. A **thermoset** is a plastic that, after having been cured by heat or other means, is substantially infusible and insoluble. Thermosets usually have a highly *cross-linked structure* in which the polymer chains are chemically interconnected with bridging groups, forming a large three-dimensional network. Because they cannot be reworked after curing, they are shaped during the cross-linking process.

Additives

Most polymers are of little practical value by themselves because of poor physical properties. For that reason, besides the polymer component, almost all plastics contain chemical additives, either to facilitate the handling or fabrication process or to produce some particular desirable property in the final product.

Some of the most common types of additive are

process additives—lubricants, mold-release agents, and blowing agents;
stabilizers—heat stabilizers, ultraviolet and visible light stabilizers, antioxidants (which can also act as process stabilizers), and antimicrobial agents, which act as preservatives by impeding the growth of microorganisms;
performance additives—inert fillers, such as glass spheres, wood flour, or chalk, used to reduce costs; reinforcing agents, such as glass fibers, organic fibers, or talcum; coloring agents, which might be dissolved dyes or dispersed pigments; impact modifiers; flame retardants; antistatic agents; and plasticizers.

There are over 4,000 different types of additive. Additives are often the crucial components of commercial products, and a large plastics additives industry grew together with the plastics industry. The global plastics additives business amounts to about $16 billion a year.

A **plasticizer** is a softening agent that is added to a polymer to facilitate processing or to enhance physical properties such as flexibility or toughness. When a plasticizer is added to the polymer material, it loosens up the structure and makes it easier for segments of the polymer chain to undergo motion. The properties of the final product are thereby changed.

Poly(vinyl chloride) (PVC) plastics in particular require large amounts of plasticizing additives, as much as 55 percent by weight. PVC accounts for almost two-thirds of plasticizer production, and phthalates are the most common PVC plasticizer. Diisononyl phthalate (DINP) is used as the softening agent in many pliable PVC toys. Di-(2-ethylhexyl) phthalate (DEHP) is used for medical tubing and intravenous bags, where the plasticizer makes up around 30 percent by weight of the bags. The environmental and health issues that have been publicly debated concerning the potential leaching of plastics additives have, more often than not, been associated with PVC.

Plasticizers are not always necessary, as is the case with ethylene-styrene copolymers, produced by random copolymerization of 25 to 80 percent styrene.

The mixing of polymers with additives is called compounding, and the result is called a *compound* (not to be confused with the more common use of that term, as in the expression *chemical compound*).

In the plastics industry, resins producers, sometimes through a broker, sell their resins either directly to a fabricator, or to a compounder, who modifies the resin for some specialized application and then sells the customized resin to a fabricator. The fabricator either processes the resin directly into finished products for end-product markets, or produces semifinished plastic products (such as plastic film) for sale to converters who manufacture the finished products (such as bags from film). A large coexisting industry provides process equipment.

Common Thermoplastics

Figure 4.3 shows some common thermoplastics and the structural formulas of the synthetic polymers from which they are processed.

> The ring structure that appears in some structural formulas stands for the phenyl group, closely related to the benzene molecule, shown in (e) in figure 4.1. One or two carbon atoms in the ring are connected to carbon atoms in the rest of the molecule. Ring carbon atoms not used to form such bonds have hydrogen atoms attached to them.

The plastics shown in figure 4.3 are commonly used for packaging. Polyethylene (PE) is produced in several forms having different densities. Density reflects the percentage of cystallinity, an important factor in determining mechanical properties. A hard form, high-density polyethylene (HDPE), contains linear chain molecules. It is used in milk jugs, laundry detergent jugs, lids, and

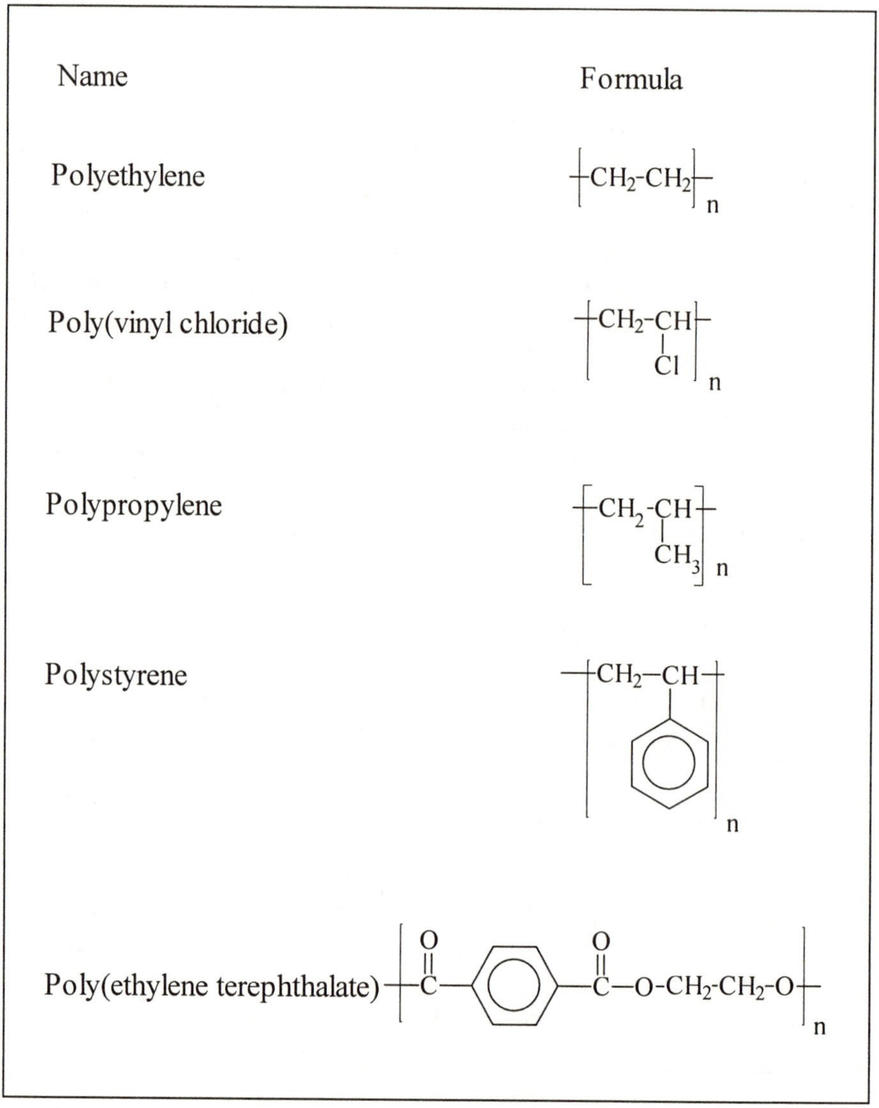

Name	Formula
Polyethylene	$\left[CH_2\text{-}CH_2 \right]_n$
Poly(vinyl chloride)	$\left[CH_2\text{-}\underset{\displaystyle Cl}{\underset{\vert}{CH}} \right]_n$
Polypropylene	$\left[CH_2\text{-}\underset{\displaystyle CH_3}{\underset{\vert}{CH}} \right]_n$
Polystyrene	$\left[CH_2-\underset{\displaystyle C_6H_5}{\underset{\vert}{CH}} \right]_n$
Poly(ethylene terephthalate)	$\left[\overset{\displaystyle O}{\overset{\Vert}{C}}-C_6H_4-\overset{\displaystyle O}{\overset{\Vert}{C}}-O\text{-}CH_2\text{-}CH_2\text{-}O \right]_n$

Figure 4.3 *Some common thermoplastics*

other containers. HDPE is used to make nonpackaging articles as well, such as sporting goods, pails, pipes, electrical insulation, and toys; its first major use, in the 1950s, was in manufacturing hula hoops.

A relatively soft form, low-density polyethylene (LDPE), contains branched chains. It is used to make food packaging, grocery bags, dry-cleaning bags, trash bags, squeeze bottles, and other packaging articles, as well as nonpackaging articles such as agricultural covers, diaper liners, and industrial sheeting. Linear low-density polyethylene (LLDPE), engineered to have shorter branch chains than LDPE, has improved properties in some respects, relative to LDPE, and competes with LDPE in generally similar markets.

Poly(vinyl chloride) (PVC) is produced in both flexible and rigid forms. Flexible PVC can be processed into clear film and sheeting and into foam, for packaging applications. Typical nonpackaging applications include floor and wall coverings, rainwear, garden hose, and electrical insulation. Some forms are popularly referred to as vinyl. Rigid PVC has major use in building, for pipes, window and door frames, and rain gutters. It is also used for credit cards and other items.

Polypropylene (PP) is used for packaging film, food-container closures, agricultural bindings, bottles, and crates for soft-drink bottles. Nonpackaging uses are wire coating, automobile battery cases, child car seats, luggage, fish nets, laboratory ware, molded parts for automobiles, appliances and other housewares, and a variety of other articles.

Polystyrene (PS) has wide application in food packaging. A clear form is used for container lids and meat trays, and for cookie, candy, pastry, and other food packages. Expanded polystyrene foam, sometimes referred to as styrofoam, is used for meat, poultry, and egg containers, take-out containers used in the fast-food market, tubs, trays, and other containers. A loose-fill packaging material is produced from polystyrene that has the desirable properties of resilience and compressibility. Nonpackaging uses include housewares, such as furniture, shower doors, flower pots, toys, combs, and disposable cups, dishes, trays, and eating utensils.

Poly(ethylene terephthalate) (PET) is used in soda bottles and boil-in-the-bag pouches. The same polymer can be processed into sheeting and used for audio and video tapes. Worldwide, approximately 12 billion pounds of PET plastics are manufactured annually. In the United States 4 billion pounds are produced, with over half being used to make bottles. It is PET's clarity, durability, and recyclability that often make it a preferred material in container applications. It was only after springwater was packaged in PET that bottled water became competitive with other soft drinks and sold widely in food markets, convenience stores, and snack counters.

Recent improvements in PET resin production and processing have allowed the manufacture of distinctive shapes that help brand recognition in packaging applications, and its use for milk products and beer will likely lead to a continued high rate of growth in production. For its use in beer bottles, barrier property problems had to be solved—carbon dioxide has to be kept in and oxygen out. One type of beer bottle has three PET layers and two barrier layers made of other plastics.

The five plastics shown in figure 4.3 account for well over 90 percent of all thermoplastics production, and almost 90 percent of *all* plastics production. Polyethylene alone accounts for approximately 40 percent of all thermoplastic resins produced—80 billion pounds a year, which corresponds to a global consumption rate of around 16 pounds a year per person. Of the 52 billion pounds of ethylene produced in the United States each year, about half is consumed for polyethylene. Poly(vinyl chloride) accounts for 20 percent of all plastics produced—44 billion pounds a year; polypropylene—19 percent, polystyrene—9 percent, and poly(ethylene terephthalate)—6 percent.

As recycling programs for plastic containers became widespread, the need to sort recycled containers became apparent. The Society of the Plastics Industry (SPI) adopted a Resin Identification Code, whereby the most common resins used for containers have been assigned numbers, and the bottoms of containers are now stamped with the corresponding number. Table 4.1 lists that coding system. For purposes of the recycling code, the abbreviation for poly(ethylene terephthalate) is PETE, not PET, and for poly(vinyl chloride) it is V, not PVC. Consumers can look on the bottom of many plastic containers and identify the resin from which the containers are made. The marking implies no guarantee of recycling, but the code nevertheless serves to encourage greater use of recycling.

Table 4.1 *Resin Identification Code*

Plastic	*Number*
Poly(ethylene terephthalate)	1
High-density polyethylene	2
Poly(vinyl chloride)	3
Low-density polyethylene	4
Polypropylene	5
Polystyrene	6
Others, and mixtures	7

Source: Society of the Plastics Industry

Today there are many other synthetic polymers that are processed into thermoplastics, but on considerably smaller scales than the *commodity* plastics shown in figure 4.3. Other *general purpose* thermoplastics include poly(vinyl acetate), poly(vinyl alcohol), and poly(vinyl butyral). Acrylics, cellulosics, polybutylene, and poly-(vinyl fluoride) are *specialty* thermoplastics.

Engineering thermoplastics are characterized by having particularly excellent mechanical or optical properties, high-temperature stability, or exceptional chemical resistance; some excel in several respects. They are often used as metal replacements in the automotive industry (e.g., for under-the-hood components), the electronics industry (e.g., for computer housings and laser printer gears), and in other applications. Established engineering plastics include polyacetals, polycarbonate, fluoropolymers, poly(phenylene sulfide), polyarylates, polysulfones, polyimides, and poly(butylene terephthalate). Combinations are also used such as acrylonitrile-butadiene-styrene (ABS) and polycarbonate-poly(butylene terephthalate).

Poly(ethylene terephthalate) qualifies both as a commodity plastic, on account of its very large sales volume and low price, and as an engineering plastic, because of its high performance properties.

Newer engineering plastics include polyetherimide, cycloolefin copolymer, aliphatic polyketones and *syndiotactic* polystyrene. In syndiotactic polystyrene the phenyl groups (see fig. 4.3) alternate from one side of the polymer chain to the other side in a regular manner, a structural feature that confers improved properties.

Common Thermosets

Thermosets account for about 10 percent of resin production. Of the thermoset resins, phenolics are the most common, making up 42 percent of the total. Others include urea resins—27 percent, unsaturated polyesters—19 percent, epoxy resins—8 percent, and melamine resins—4 percent.

In phenolic resins, two low-molecular-weight compounds—such as phenol and formaldehyde—are partially polymerized to form a low-viscosity liquid, which is then mixed with additives and processed using methods such as compression molding, transfer molding, or casting. Fillers such as wood flour, cotton flock, mica, or glass fibers are added, sometimes making up well over half the final weight. The amount depends on the strength characteristics and cost desired in the final product. In resin transfer molding, the dry reinforcement is placed in the mold before the mold is closed and resin admitted. The final curing step is typically at a high temperature and usually at high pressure, during which cross-linking proceeds to form an unfusible "set."

Phenolics are resistant to heat and chemicals, and are used for circuit boards, plywood adhesive, fiberglass binder, and a wide variety of molded products, large and small. Thermosets can provide very sturdy materials and are widely used in the furniture, transportation, and construction industries. Because they cannot be reheated and reworked, thermosets are usually not readily recyclable.

Biodegradable Synthetics

Some synthetic **biodegradable** polymers (See Chapter 5) are shown in figure 4.4. Poly(vinyl alcohol) (PVA) is prepared from the polymerization of vinyl acetate followed by alcoholysis. It is water soluble and is used extensively to produce films, fibers, paper coatings, and adhesives. The name poly(vinyl alcohol) is based on the repeating structural unit that appears in the polymer; it is not monomer based. (There is no stable "vinyl alcohol" species.)

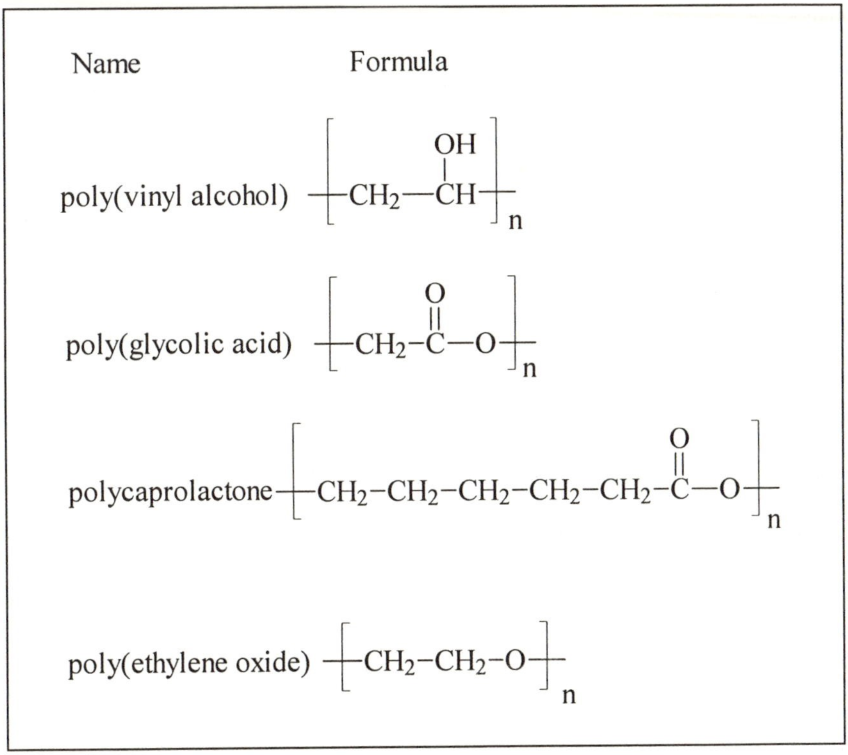

Name	Formula
poly(vinyl alcohol)	$\left[-CH_2-CH(OH)- \right]_n$
poly(glycolic acid)	$\left[-CH_2-C(=O)-O- \right]_n$
polycaprolactone	$\left[-CH_2-CH_2-CH_2-CH_2-CH_2-C(=O)-O- \right]_n$
poly(ethylene oxide)	$\left[-CH_2-CH_2-O- \right]_n$

Figure 4.4 *Biodegradable synthetic polymers*

Poly(glycolic acid) is a polyester, and it is thermoplastic. It is synthesized from condensation polymerization of glycolic acid, or catalytic-ring-opening polymerization of the glycolide. The monomer, glycolic acid, is a naturally occurring compound, but it is generally produced commercially through chemical synthesis. Poly(glycolic acid) is used in a variety of pharmaceutical and biomedical applications, including drug delivery systems, wound treatment applications, and implants.

In drug delivery systems, therapeutic agents are released from microcapsules, microspheres, or nanocapsules by using time controlled degradation of the polymers in the encapsulating layer. The capsules can either be pressed into a pill and administered orally, or injected subcutaneously. In veterinary medicine the polymers form the matrix of large pills—boluses—which deliver medicine in a preset time interval in, for example, the rumen of cattle. Implants and films are also used for drug delivery. Many types of bioactive compounds can be encapsulated, including antibodies, hormones, and antitumor agents.

In wound treatment applications, the plastic is made into sheets or spun from solution to form gauze. The sheet or gauze can then be used externally to promote wound healing, or internally to separate tissue during healing. Small pieces of fiber left in the wound are not a cause for concern because they will degrade. Resorbable surgical sutures can similarly be made so as to be absorbed by the body at a slow, predetermined rate. A degradable plastic can also be used as a temporary scaffold for new tissue growth, whereby it is replaced by natural tissue during the healing process. Surgical implants are used to join body parts, where the material degrades as the new natural material is regenerated.

For biomedical applications the polymer has to be *biocompatible*—nontoxic and not rejected by the intended organism. For some types of biomedical application a material might be required to be biocompatible and nondegradable. For other types of application a material might be required to be biocompatible and degradable, or partially resorbable by the organism. A poly(glycolic acid) implant, for example, will degrade in a time determined by its size and shape, but generally in less than twelve months.

Polycaprolactone is synthesized by ring-opening polymerization of caprolactone. Like poly(glycolic acid), it is a polyester and thermoplastic. It is generally used to modify the properties of other degradable plastics for specific needs. It has agricultural applications as a matrix system for the controlled release of pesticides, herbicides, and fertilizer, and it is used in blown films for degradable compost bags for food and yard wastes.

Poly(ethylene oxide), structurally equivalent to poly(ethylene glycol), is water soluble and thermoplastic, and is used in adhesives, lubricants, cosmetics, pharmaceuticals, antifreeze agents, printing inks, paper coatings, and other products.

The degradability of these polymers can be attributed to their chemical structures. Poly(vinyl alcohol) has a backbone of carbon-carbon single bonds, which generally indicates poor degradability. But in poly(vinyl alcohol) the substituent –OH groups on alternating carbon atoms makes for strong interactions with water; the polymer is **hydrophilic**. The polymer is thereby extremely soluble, in contrast to the four polyolefins in figure 4.3, which are **hydrophobic** and insoluble.

The remaining polymers in figure 4.4 are polyesters or a polyether, with oxygen atoms in the polymer backbone. It is the appearance of a noncarbon atom (a heteroatom) in the backbone that constitutes the single most important reason for the degradability of the polymers. (Polymer degradation is described in more detail in Chapter 5.)

Fibers and Elastomers

Polymers are also used to manufacture fibers, elastomers (rubbers), adhesives, and paints and other coatings—in some cases the distinctions are not clear-cut.

Fibers are formed from polymers by spinning. In one spinning process the polymer is dissolved to form a viscous solution. The solution is then pumped to a spinning machine where it is forced through a spinerette—a metal plate or disk perforated with many holes; the holes may be as small as 0.001 inch or as large as 0.010 inch in diameter. In dry spinning the solution passes through a column of hot gases that evaporates the solvent and solidifies the polymer into filaments. In wet spinning the solvent is leached out by another solvent. In either case the filaments may undergo further processing (lubrication, crimping, cutting) depending on the specific polymer and its intended end-use application. It is finally packaged in a form that is also dictated by the application.

In other spinning processes a dispersion might be used rather than a solution, and if the polymer is stable at its melting point, melt spinning is feasible.

Some of the same synthetic polymers used to make plastics are also used to make fibers—high-density *oriented* polyethylene fibers have an extraordinary strength-to-weight ratio and are used for space tethers. Synthetic fibers are often made into cloth, as when poly(ethylene terephthalate) fibers are used to manufacture polyester permanent-press fabrics, home furnishings, and sails. Globally, fibers are the dominant application of poly(ethylene terephthalate) resin; it is strong, easy to dye, and shrink-resistant. It can also be extruded from melts to form thread.

Nylon, a polyamide, was developed in the 1930s as the first synthetic polymeric fiber. Nylon—smooth, resilient, lustrous, and easy to wash—is used to make carpeting, tents, sails, rope, and other articles. Acrylics—lightweight and quick drying—are used in baby clothes, sweaters, socks, carpeting, and craft yarn. As with plastics, there are many different synthetic fibers having a wide range of properties.

The global synthetic-fiber market, of around 60 billion pounds a year, is dominated by polyesters—40 billion pounds a year; polyamides—10 billion pounds a year; and acrylics—6 billion pounds per year.

Elastomers, or **rubbers**, are polymer materials that can be deformed by an applied stress, and then recover their original form when the stress is removed (as long as their elastic limit has not been exceeded). In 1770 the British chemist Joseph Priestley introduced the term *rubber* after he observed that the natural milky juice of certain tropical plants could be coagulated and used to rub out pencil marks. Synthetic rubbers are either general-purpose rubbers—used mainly for tires—or specialty rubbers—important for a variety of other uses. Two major types of rubber are copolymers of styrene and butadiene (SBR rubbers) and copolymers of acrylonitrile and butadiene (NBR rubbers). Other types include neoprene and polyurethane rubbers, polyacrylate elastomers, silicone elastomers, and fluoro-elastomers. Composites and blends are used as well, such as a mix of SBR rubber and polybutadiene that is used for tire treads.

The physical properties of elastomers in their pure form are too poor to serve any practical purpose. As with plastics, additives give the final materials the strength, hardness, wear resistance, and other properties that make them useful. In tires, for example, reinforcing fillers—mainly carbon black—improve tensile strength, abrasion resistance, and tearing strength, without introducing serious deterioration of the elastic properties that keep rolling resistance low. For nonblack applications, silicas or silicates serve as reinforcing fillers. Nonreinforcing fillers—like chalk or mica—are also often added to rubbers to reduce the price of the material.

4. Chemical Nature of Plastics

The number of possible polymer materials—plastics, fibers, and elastomers—is enormous because polymers have a wide range of chemical structures, they can be combined in composites or blends, the range of additive formulations is large, and their processing can be varied. New polymer materials are constantly being found—continuously expanding the frontier.

In a search for new, environment friendly plastics, we can go beyond the world of synthetic polymers, and explore the world of *natural polymers*—a world well studied for other purposes but now being probed with the goal of finding novel plastics. First, however, if benign degradation is also a goal, degradation itself has to be scrutinized in more detail so that the different behaviors of synthetic and natural polymers can be appreciated.

5

Plastics Degradation

Plastics after Use—An Introduction

The environmental degradation of plastics is not a simple process; the details of how plastics degrade after disposal are not thoroughly understood and are the subject of current research. A useful starting point in considering plastics degradation is to picture three general, idealized categories of plastics according to how readily they degrade in a specified environment.

(Only environmental degradation is considered here. Degradation under extreme conditions, such as incineration, is not included.)

Nondegradable Plastics

Until recently there has been no interest in making commodity plastics that are anything but extremely stable in almost every environment. Virtually the total efforts of polymer chemists and plastics engineers have been directed toward increasing resistance to all types of degradation, and those efforts have been highly successful. Commodity plastics are typically stable in almost all environments. In some environments, objects made from them remain intact for many years.

Their persistence, for the most part, originates in three of their properties that make them so useful for many applications. They are generally strong. They are water resistant. And microorganisms do not readily attack them.

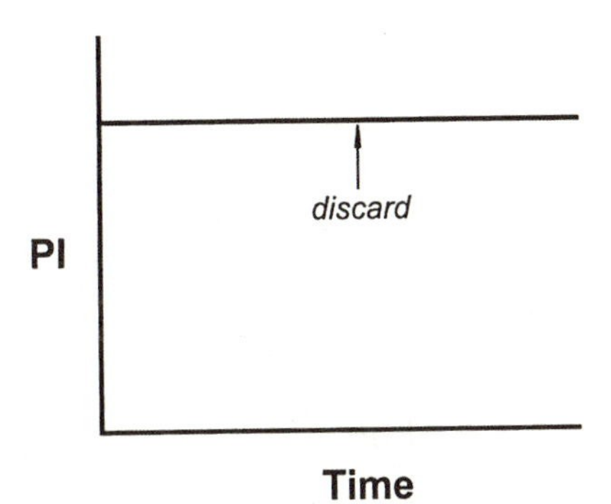

Figure 5.1 *Degrade curve of a nondegradable plastic*

In the simplest of approximations, plastics degradation can be represented pictorially, as in figure 5.1. Time is indicated on the horizontal axis. The vertical axis bears the label PI, for "Performance-Integrity" index. The PI index indicates how physically intact the plastic is, and how resistant it is to degradation. As the plastic degrades the PI index drops. When the plastic has degraded completely, its PI index reaches zero.

Figure 5.1 shows the degrade curve of a plastic that does not measurably degrade on any reasonable time scale. It approximates many current commodity plastics, including the polyolefins—polyethylene, poly(vinyl chloride), polypropylene, and polystyrene—and even the polyester poly(ethylene terephthalate) (fig. 4.3). During the useful lifetime of the plastic, the PI index remains constant; it is a sturdy plastic. But after its useful lifetime, it persists with little or no degradation. A discarded plastic sandwich bag may be around long after the sandwich is gone and its consumer deceased!

Readily Degradable Plastics

A readily degradable plastic, after its useful life has ended, simply "self-destructs," like the tapes in television's and films' *Mission Impossible* stories. After the time required for useful service, during which it retains all the properties that it was formulated and processed to have, it simply falls apart and is assimilated by the

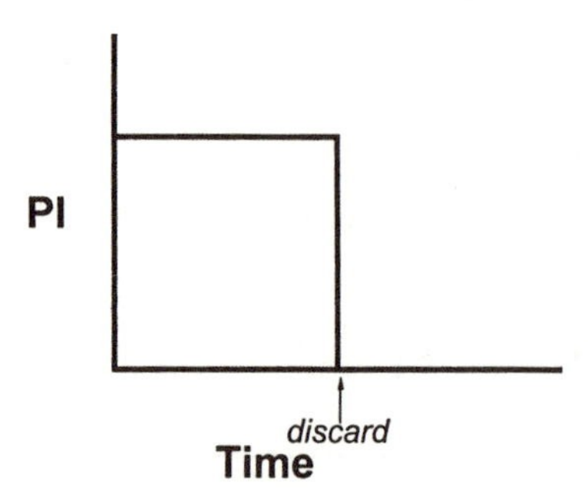

Figure 5.2 *Degrade curve of a readily degradable plastic*

pervasive microorganisms found throughout nature. It returns to the ecosystem in an environmentally harmless manner.

The degrade curve of an idealized, readily degradable plastic is shown in figure 5.2. Initially the PI index is maintained at a constant level. Then, after a period of useful life, it is discarded and completely degrades. Moreover, it does so rapidly, whereupon its components are returned to the ecosystem.

Some synthetic polymers derived from fossil resources approximate such behavior, like poly(vinyl alcohol), poly(glycolic acid), polycaprolactone, and poly(ethylene oxide) (fig. 4.4). They are generally specialty polymers, rather than commodity polymers. They are used in detergents, water treatment, thickeners, and other applications where rapid degradation after use is an important requirement of the application. Their use in detergents, for example, aids wastewater treatment. The actual time scale for their degradation varies with the application and the disposal environment but typically might be on the order of days or a few weeks.

Programmed Degradable Plastics

Programmed degradation, or *controlled* degradation, is a relatively new idea in which the goal is to program plastics to degrade in a predetermined time under specific conditions according to the needs of particular applications. It has only been in the last twenty or thirty years that producing plastics with controlled lifetimes has been attempted—and achieved.

One approach has been aimed at the litter problem where exposure to natural sunlight is common. The objective is to modify the resin so as to promote **photodegradation**, degradation that results from the action of natural sunlight. The strategy is to attach a photosensitizing group to the polymer chains by chemical means. When the photosensitive group is exposed to natural sunlight, it absorbs radiation, which causes the chain to break and form smaller segments, a process called chain *scission*. As photodegradation proceeds, the chains are broken in more and more places, and the plastic litter is destabilized through embrittlement. Eventually it becomes fragile and fragments. Erosion by wind and rain completes the breakdown of the embrittled plastic into a friable powder.

Resins of otherwise nondegradable commodity polymers, such as polyethylene, polypropylene, and polystyrene, can thereby be modified to make them photosensitive. The sensitizing group, such as a ketone group, is introduced into the polymer chain by copolymerization with an appropriate ketone monomer. Or, carbon monoxide (CO), for example, can be copolymerized with ethylene to produce a photodegradable polymer. The copolymerization of ethylene and carbon monoxide is mainly used for the large-scale manufacture of photodegradable six-pack holders for beverage containers. Compositions containing 1 percent CO will photodegrade after about three weeks' exposure to outdoor sunlight and break up into small particles.

Sometimes a photosensitizing group can be incorporated after polymerization. Metallic salts, of nickel, cobalt, iron, or some other metal, are incorporated as photosensitive additives during processing.

Delayed action is sometimes required to provide an initial period of stability during processing, shelf life, and use, followed by degradation in some particular, definite amount of time. Even that highly specific type of behavior can be achieved, as when metal ion complexes are added. The complexes act as stabilizers during processing but then decompose, after an induction period, in a controllable manner to form products that are photoactivators. By a careful choice of the right combination of stabilizer and activator concentrations, the length of the induction period and the rate of the photodegradation that follows can be controlled.

Newer technologies have been developed that do not rely on photodegradation, so that exposure to natural sunlight is not necessary. There are now additive formulations that, when compounded with a polyolefin like polyethylene or polypropylene, promote environmental degradation through the chemical processes of **oxidation** and **hydrolysis**.

Oxidation is, generally, a chemical reaction in which electrons are lost by an atom or molecule. It can be carried out either by direct combination with oxygen or by reaction with chemicals (oxidizing agents) that produce the same fundamental type of chemical change—a transfer of electrons from the atom or molecule being oxidized to the oxidizing agent. In reactions of organic molecules, oxidation is reflected in the addition of oxygen atoms to a molecule or the removal of hydrogen atoms. For example, alcohols can be oxidized to aldehydes (in the case of primary alcohols) or ketones (from secondary alcohols) and then further oxidized to carboxylic acids. The oxidation of ethanol to acetaldehyde, then to acetic acid provides for elimination of alcohol by the human body.

$$CH_3CH_2OH + \text{oxidizing agent} \rightarrow CH_3CHO$$

$$CH_3CHO + \text{oxidizing agent} \rightarrow CH_3COOH$$

In the direct air oxidation process, oxygen is the oxidizing agent and the process produces H_2O. *Epoxidation* is the formation of epoxides, often from alkenes.

Hydrolysis is any chemical reaction in which a compound is converted into another compound by taking up the elements of water. Esters, for example, undergo hydrolysis to produce the corresponding carboxylic acid and alcohol; the net reaction is given by

$$RCO_2R' + H_2O \rightarrow RCO_2H + R'OH$$

A specific example is the hydrolysis of methyl acetate to produce acetic acid and methyl alcohol.

$$CH_3CO_2CH_3 + H_2O \rightarrow CH_3CO_2H + CH_3OH$$

Oxidative degradation is degradation that results from oxidation. The newer polyolefin additive formulations allow oxidative degradation to be initiated either by natural daylight, or by heat, or even by mechanical stress. The intended applications for the resins are plastic products where disposal might include earth burial, as with compost bags and agricultural mulch covers.

Moreover, the fragments formed by oxidative degradation are *wettable,* leading to increased interactions with water and promoting *hydrolyis*. *Hydrolytic degradation* is degradation that results from hydrolysis. Through oxidative and hydrolytic degradation the polyolefin chains undergo progressive chain scission. In time, embrittlement sets in and fragmentation becomes extensive.

Fragmentation, although an essential element of degradation, is not the same as total degradation. A plastic may fragment after being discarded, but may not further readily degrade so as to be decomposed, for example, by microorganisms. Through embrittlement a large piece of plastic becomes fragile, and the many small pieces of plastic eventually turn into a friable powder, perhaps even invisible to the naked eye.

In theory, fragmentation of the polymer chains in the plastic may make the plastic more susceptible to other modes of degradation, including biodegradation. In practice, questions related to the environmental fate and ecological effect of polyolefin fragments have even today not been conclusively resolved and make up an active area of current research.

For example, if agricultural covers made from modified polyethylene resins are plowed under at the end of a growing season, how much of the cover remains as intermediate-length polyethylene chains by the time a new growing season begins, or after two successive years of use? Most important, what is the long-term effect of *any* accumulating residues on agricultural productivity? It may be that there are no harmful effects, but the answers to such questions are not easy to find with short-term studies, and few long-term studies have been done. Because long-term agricultural-productivity is at stake, and because the need for future corrective cleanup operations has to be avoided, the questions are important.

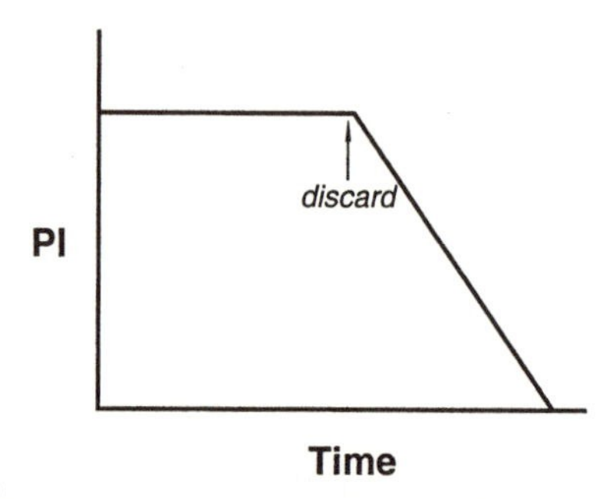

Figure 5.3 *Degrade curve of a programmed degradable plastic*

The idea of programmed degradation of plastics goes beyond any particular type of plastics material and any particular application. If figures 5.1 and 5.2 are taken to represent two opposite limiting cases of degradation, one can imagine searching for intermediate programmed degradation behaviors, to match any specific practical application and various disposal methods.

What is desirable in a programmed degradable plastic is to have adequate performance properties initially, and no significant decrease in performance properties during the planned useful lifetime (perhaps including several reuses). On the other hand, after the period of use, degradation is to begin upon disposal, starting with fragmentation or surface erosion. Ideally each plastic item could have a label: "This plastic will self-destruct in (two weeks, six months, one year, ten years), when disposed of in (soil, compost, seawater, etc.)." In some applications, such as compost bags and agricultural covers, degradation is intended to begin when the material is put into use.

Figure 5.3 illustrates the degradation of one type of programmed degradable plastic. In particular applications it might be desirable, and possible, to have fragmentation followed by a relatively rapid total degradation, on a time scale of perhaps weeks to months or a year.

In other cases (fig. 5.4) fragmentation and partial degradation might be followed by a slower approach to total degradation, with

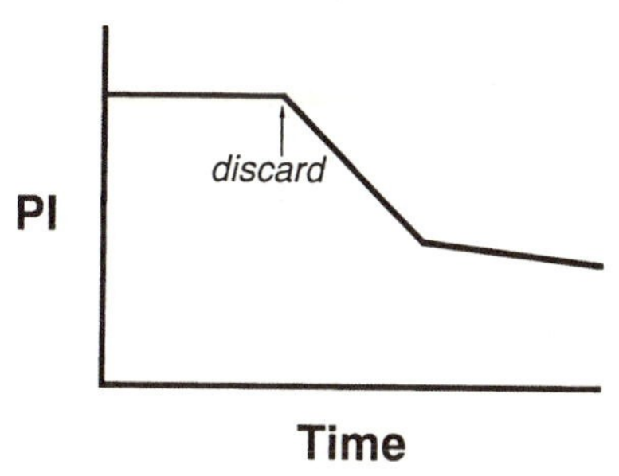

Figure 5.4 *Degrade curve of a programmed degradable plastic degrading in two stages*

perhaps no harmful effects on the environment. Modified polyolefin resins are likely better described by a degrade curve similar to figure 5.4 in which the time scale of the second stage of degradation has not yet been ascertained.

Degrade curves do not convey the complexity of plastics degradation; they are simple cartoons. Plastics degradation involves a series of processes with many variables. It entails the workings of the fundamentals of thermodynamics and, especially, kinetics.

Thermodynamics and Kinetics

According to the laws of **thermodynamics**, many compounds, including plastics, have a tendency to degrade. A consequence of the Second Law of Thermodynamics is that the universe is constantly tending to become more and more disordered; **entropy** is a measure of that disorder. Events that occur spontaneously increase the entropy of the universe, and vice versa. But in thermodynamics the word *spontaneously* does not mean the event will take place fast; it only means that the event will happen without requiring any work or outside effort. Ice exposed to a temperature of 20 °C and a pressure of one atmosphere will melt because the entropy of the universe thereby increases. Likewise, plastics have a tendency to degrade because the universe's entropy increases when they do.

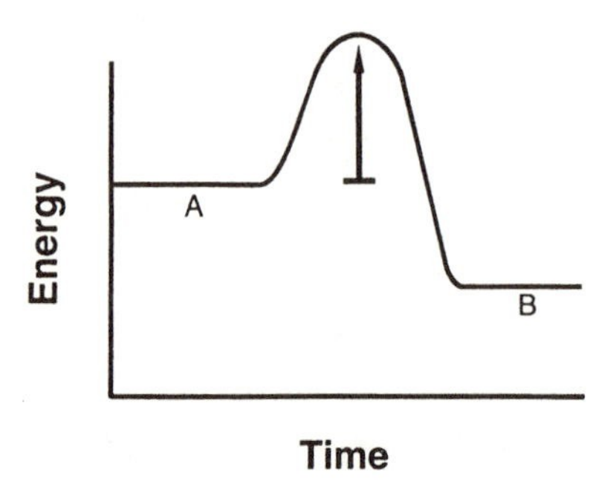

Figure 5.5 *The course of a chemical reaction*

The Second Law of Thermodynamics does not tell us how long it will take for reactions to take place. Some reactions take place in much less than a second, while others take minutes, hours, or years. A chemical reaction rate is the measure of how fast a chemical reaction takes place, and the branch of chemistry that deals with chemical reaction rates is chemical **kinetics**.

Thermodynamics and kinetics describe two different aspects of chemical reactions, including the degradation of plastics. Figure 5.5 illustrates the relationship between thermodynamics and kinetics for one of the simplest pictures of chemical kinetics—*collision theory*. The progress of the reaction is indicated along the horizontal axis; it is labeled the time axis in figure 5.5, but is more accurately called the *reaction coordinate.* The vertical axis is used to indicate the amount of energy in the chemicals taking part in the reaction.

The energy curve starts at the left with the reactant (or reactants) having an energy A. At the right the reaction has completed, producing the product (or products) with an energy B. If the energy of the product is lower than the energy of the reactant, as in figure 5.5, the reaction is *exothermic* and energy is released as heat. If the energy of the product is greater than the energy of the reactant, the reaction absorbs energy as heat and is *endothermic.* The net change in energy on going from reactant to product is a thermodynamic property of the reaction. Collision theory aims to explain what factors determine the *rate* at which the reaction proceeds from the left to the right along the reaction coordinate.

Reactant molecules collide in the course of their random motion, and in collision theory one of the factors that determines the reaction rate is the frequency with which reacting molecules collide—the encounter rate. The encounter rate itself is determined by several factors, including the concentrations of the various reactant molecules: the higher their concentrations, the more likely molecules are to encounter one another in the course of their constant motion. Conversely, if a required chemical is present in too small a concentration, the reaction may become immeasurably slow. The reaction temperature also affects the encounter rate: molecular velocities increase with temperature, leading to more frequent collisions. The sizes of the reacting molecules are also relevant: large molecules collide more frequently than small molecules.

Encounter rate is not the only factor that determines the reaction rate because not all encounters lead to chemical changes. In order to get from A to B an energy barrier has to be overcome; it is indicated in figure 5.5 by the arrow. If the molecules have sufficient kinetic energy (energy of motion) the collision results in the chemical bonds of reactant molecules being broken and the new chemical bonds of the products being formed. The energy barrier height plays a role in the kinetics of the reaction because the higher the barrier, the smaller the proportion of reactant molecules having the required energy, and therefore the smaller the proportion of encounters leading to reactants.

At higher temperatures the average kinetic energy of the molecules is increased, increasing the proportion of reactant molecules having sufficient energy to surmount the energy barrier. This provides a second role for the temperature in determining chemical reaction rates (the temperature also determines the encounter rate by affecting molecular velocities).

A third factor playing a role in determining reaction rates is the orientation of the reactant molecules as they approach one another in collisions. The barrier height may depend on orientation, and two molecules may not be as predisposed to react along one line of approach as they are along another. The one-dimensional representation of figure 5.5 is simplified. A more accurate picture requires considering the energy as a function of all relative approach orientations for all molecules taking part in a reaction.

Collision theory provides a rather simple picture of chemical kinetics, but it does a reasonably good job in accounting for the observed features of many chemical reactions, especially for reactions occurring in the gas phase, but also for some reactions that take place in solution or even in solids. In particular it describes the observed temperature dependence of many chemical reaction rates: from experimental data an empirical *activation energy* is extracted and interpreted as representing the energy barrier that appears in the theory.

Transition state theory is a modification of collision theory intended to be applicable to more general types of chemical reaction, including those that take place in solution where collisions with solvent molecules also occur. The peak in the energy curve is given particular significance, and is said to correspond to the formation of a *transition state.* The peak of the energy curve represents the climax of the chemical reaction, when chemical bonds in reactant molecules are being broken and being replaced with the bonds of the product molecules. From the transition state the reaction can either proceed farther to the right, where the energy decreases to that of the product (or products), or revert to the initial reactants. Transition state theory is widely used in interpreting chemical kinetics data, but the simpler collision theory introduces all of the notions required for some understanding of the kinetics of plastics degradation, particularly the dependence of degradation rate on *concentrations, temperature,* and *chemical nature of the plastic.*

The energy curve in figure 5.5 is simplified also in its implication that a chemical reaction takes place in one step, with a single energy maximum separating reactants and products. A chemical reaction usually occurs in more than one step. For example, ozone (O_3) might be pictured as decomposing into oxygen (O_2) following a single collision whereby two ozone molecules produce three oxygen molecules, giving the observed overall result: $2\ O_3 \rightarrow 3\ O_2$. The situation in the real world (in this case, the real atmosphere) is more complicated, and in most cases a great deal more complicated. One alternative, for example, is that an ozone molecule absorbs energy from sunlight to produce an oxygen molecule and an oxygen atom: $O_3 \rightarrow O_2 + O$, followed by a collision between that oxygen atom and another ozone molecule to produce two oxygen molecules: $O + O_3 \rightarrow 2\ O_2$. The net result is the same as in the first description (the production of three oxygen molecules from two ozone molecules), but the two-step process gives a fundamentally different picture of what occurs on a microscopic level. In the presence of sunlight and with the commonly low concentration of ozone, the latter sequence of individual *elementary reactions* is more probable than the single-collision description. The particular sequence of elementary reactions that describes the *microscopic reaction path* of a chemical reaction is called the *reaction mechanism* of the chemical reaction.

When there are several steps in a reaction mechanism, there may be more than one energy barrier to be overcome, but often one of the energy barriers is significantly higher than the others, which results in one of the steps of the mechanism being the *rate-limiting* step. Consider an analogy with a single-lane highway, where the traffic cannot move faster than the slowest vehicle—the rate-limiting vehicle.

Further complicating chemical kinetics in real-world situations is that a chemical reactant may be able to take part in several different reactions, along several reaction paths, with each path producing a different product. For example, a molecule might be able to undergo both an oxidation reaction and a hydrolysis reaction. Depending on the conditions and chemical environment, one reaction path might be overwhelmingly favored over another; in different circumstances both reaction paths might be accessible simultaneously, producing a mixture of oxidation and hydrolysis products. Each reaction is controlled by its own characteristic kinetics, including its own characteristic energy barrier height.

Much of applied chemistry deals with the challenge of increasing the rates of desirable chemical reactions through the use of **catalysts**. A catalyst is a substance that accelerates a chemical reaction by enlarging the range of possible reaction paths by which the reaction can proceed, thereby permitting a path with a lower barrier height than that of the uncatalyzed reaction. The result is a faster reaction rate at the same temperature. The catalyst takes part in the reaction but undergoes no net chemical change; it is not itself consumed in that reaction. (Although there is no net chemical change in the catalyst from the catalyzed reaction, there is often some small loss of catalyst through *other* reactions that can occur at the same time.)

Catalysts can be just about anything. Acids often catalyze reactions; metals sometimes make good catalysts. At times ordinary light promotes chemical reactions that otherwise take place much more slowly.

Biodegradation—Nature's Recycling

Light absorption by chlorophyll initiates **photosynthesis**, the synthesis of carbohydrates from carbon dioxide and water. Through photosynthesis, plants capture energy from the sun and reorganize the atoms of carbon dioxide and water molecules into the chemical machinery that makes plant life possible. Animals, incapable of photosynthesis, evolved by learning how to use the energy stored in plants. The chemistry that occurs in these living organisms—biochemistry—is very efficient; energy waste is at a minimum. That efficiency is the result of millions of years of evolution during which *enzymes* developed in both animals and plants.

Enzymes are protein molecules that catalyze biochemical reactions (the substance whose reaction is being catalyzed is called the *substrate).* Enzymes are exceedingly effective catalysts; the enzyme saccharase increases the rate of the hydrolysis of sucrose by a factor of 10^{12} compared to its nonenzymatic acid hydrolysis. In animals some enzymes have the specialized role of degrading food. If food is not chewed, enzymes in the mouth, excreted by salivary glands, do not have a chance to get digestion started properly and more work is left for enzymes in the stomach, in some cases resulting in indigestion.

Enzymes are also found in **microorganisms**, including **bacteria**, **fungi**, and **algae**. Microorganisms contain enzymes that can induce chemical degradation and are responsible for much of the chemical degradation that occurs in nature. Cows cannot digest grass because the enzymes that degrade grass are not part of the cow's biochemical system. The enzymes responsible for the digestion of grass are found in microorganisms that reside symbiotically inside the cow.

Enzymes not only play a critical role in sustaining living organisms; they also play an important role as degradation catalysts that help maintain important balances in nature as life forms proceed from generation to generation.

Biodegradation, or **biotic degradation**, is chemical degradation brought about by the action of naturally occurring microorganisms such as bacteria, fungi, and algae. (Chemical degradation that does not involve biological activity is **abiotic degradation**.) As biodegradation proceeds it produces carbon dioxide and/or methane. If oxygen is present the degradation that occurs is **aerobic degradation**, and carbon dioxide results. If there is no oxygen the degradation is **anaerobic degradation**, and methane is produced instead of carbon dioxide. Under some circumstances both gases are produced.

Mineralization is defined as the conversion of biomass to gases (like carbon dioxide, methane, and nitrogen compounds), water, salts and minerals, and residual biomass. Mineralization is complete when all the biomass is consumed and all the carbon in it is converted to carbon dioxide. Complete mineralization represents the reentering of all chemical elements into natural biogeochemical cycles. Sometimes *mineralization* is used to be synonymous with *complete mineralization,* but the distinction is an important one.

Under some conditions 100 percent of organic matter may mineralize over a relatively short time period—perhaps a year or less. Even in nature, however, organic matter may become incorporated into materials—like peat—that act as carbon reservoirs, and the organic matter then mineralizes only over very long periods of time.

It is through the action of microorganisms in soil that organic matter is degraded to the point that it becomes *humus*. There may be billions of bacteria in a teaspoon of topsoil (about 5 milliliters) and tens of thousands of different species. Microorganisms give life to soils and make soils living systems; without them the soil would be sterile. In addition to their functions of decomposing organic matter and recycling nutrients, microorganisms are important for the fixation of nitrogen, the maintenance of soil structure, and the detoxification of pollutants. (In composting, the aim is to enhance all of those functions.)

Bacteria start the process of decaying organic matter, but are soon joined by fungi and protozoa. **Macroorganisms**, visibly large organisms, also contribute to the degradation of organic matter. In soil, innumerable insects (centipedes, millipedes, and beetles) and invertebrates (like snails and slugs) feed on microorganisms and plant residues and help to produce humus. Larger organisms, like the very important earthworm, and even larger animals, also take part in the process, through ingestion, mastication, and excretion.

The rate at which biodegradation occurs in soil depends on soil conditions such as temperature, moisture level (a measure of the concentration of water), degree of aeration (a measure of the concentration of oxygen), acidity (a measure of the concentration of acids), and the concentration of microorganisms themselves. Under extremely unfavorable conditions degradation rates can be reduced to nearly zero.

Low temperatures strongly inhibit degradation. In very cold climates human remains have regularly been discovered that are thousands of years old, and yet are amazingly well preserved on account of the low temperature. In 1991 a 5,200-year-old, but nevertheless well-preserved, human body was found in a glacier in the Ötztaler Alps. The extremely low temperature had brought normal decay to a halt. Through detailed scientific investigations of the mummy (now called Ötzi), we have been given a fascinating new view of life in the Neolithic era.

Moisture is also important; it supports hydrolytic degradation. Newspapers, even though they are potentially biodegradable when sufficient moisture is provided, will not environmentally biodegrade in a landfill if moisture is inadequate. Twenty-year-old hot dogs have been found undegraded in landfills, probably on account of inadequate moisture.

Aeration supports oxidative degradation, and the degree of aeration determines whether aerobic or anaerobic degradation—or both—takes place. Although there are many bacteria that thrive on an oxygen-free environment, there are many more that use oxygen.

Degradation also requires that the soil be **microbially active**. Degradation rates can be reduced to nearly zero in a sterile environment, or when the concentration of microorganisms is very low. The famous "Tollund Man," discovered in 1950 in a peat bog near Tollund, Denmark, turned out to be more than 2,000 years old, preserved when the acidic water of the peat bog prevented the growth of decomposing bacteria. Tannins in the bog had also turned the skin to leather.

Microorganisms help decompose organic matter in marine environments as well. There may be millions of microorganisms in one milliliter of seawater or in one milligram of sediment.

Through natural decomposition, carbon atoms are thereby returned to the ecosystem and once again made available for living plants to recapture them through the process of photosynthesis. Animals, through their intake of food, through respiration, through excretion, and through their ultimate decomposition, are also part of the balanced exchange of carbon. These processes are part of the **carbon cycle**, through which carbon atoms are continually recycled within the ecosystem from one molecule to another. Nature's way of recycling organic matter by means of the carbon cycle is simple and efficient.

In addition to the carbon cycle, the *grand cycles of nature* include the nitrogen cycle, sulfur cycle, phosphate cycle, and hydrogen cycle (strongly coupled to the carbon cycle). The carbon cycle is particularly relevant to plastics, but all cycles are involved in biodegradation.

The total amount of carbon dioxide in the atmosphere is kept in balance through nature's cycles. As biomass is produced through photosynthesis, carbon dioxide is consumed; as biomass degrades, carbon dioxide is released. The large use of fossil fuels for energy over a relatively short period of time has raised concerns as to whether the natural mechanisms for maintaining carbon dioxide balance have been strained. By one estimate the carbon dioxide flow from land into the atmosphere (7 billion tons of carbon per year) has been increasing at an annual rate of approximately 10 percent of the baseline preindustrial flow. The increase has contributed to a buildup of atmospheric carbon dioxide of approximately 0.4 percent a year.

Carbon dioxide accounts for about half the **greenhouse effect** by which some atmospheric gases absorb and reradiate long-wavelength (infrared) radiation. A *natural* greenhouse effect is responsible for increasing the annual global mean temperature of Earth's atmosphere and surface from a rather chilly –18 °C, without the effect, to 15 °C, which most people find more comfortable. Whether an *enhanced* greenhouse effect, from the buildup of greenhouse gases, is leading to a significant additional warming of the planet is of crucial environmental importance, and the subject of much current research.

Biodegradation and abiotic degradation both play roles in plastics degradation.

Degradation of Plastics

Degradation of a plastic is defined as a deleterious change in its appearance, physical properties, or chemical structure. Describing plastics degradation, measuring it, and controlling it are all complicated by three major factors.

(1) Plastics can, and do, degrade by many routes, consecutively or simultaneously

The plastic can be fragmented through physical forces. Fragmentation often plays an important role in the early stages of degradation, and can be brought about by physical forces of a mechanical nature, as when a plastic is cracked, broken, shredded, pulverized, or worn through surface erosion. Physical deterioration can also come about through interactions with water or other solvents whereby leaching occurs. The degree of fragmentation is important, because when there are a large number of small pieces the total surface area of plastic that is exposed for chemical reaction is greater; the "concentration" of accessible plastic material is increased.

Chemical changes within the plastic can occur and may begin with abiotic degradation. Chemical degradation occurs through reactions of the plastic with chemicals in the surroundings, such as water, acids, or other chemicals. Degradation brought about by chemical reactions generally involves chain scission—fragmentation of the polymer chains in the plastic. Hydrolysis and oxidation can both be important in the case of plastics. Surface erosion, for example, can be the result of chain scission resulting from chemical hydrolysis. Photodegradation, a result of ordinary daylight, can also occur and it too results in chemical changes.

At some point the plastic may be attacked effectively by microorganisms—the onset of biodegradation. Biodegradation is generally considered as consisting of both enzyme-catalyzed hydrolysis and nonenzymatic hydrolysis. Enzymatic degradation can be carried out either by extracellular enzymes present in the microorganism's environment or by intracellular enzymes. Both result in chain scission whereby the polymer chains are cleaved into smaller segments.

The enzymes may be either endoenzymes, which cleave internal linkages within the chain, or exoenzymes, which cleave terminal monomer units sequentially. Endoenzymes cleave the internal chain linkages randomly, which results in a rapid decrease in molecular weight; the sequential cleavage of terminal segments leads to less dramatic immediate changes in molecular weight.

When the initial fragmentation is extracellular, polymer fragments may at some point become small enough to be transported into the cell, where degradation continues to the point of complete mineralization. Mineralization may be carried out by the same organism that provided the extracellular enzymes for the initial fragmentation or by different organisms.

Under some conditions macroorganisms contribute to the degradation of plastics through ingestion, mastication, and excretion.

All of these pathways are potential routes for plastics degradation.

(2) How plastics degradation proceeds in a specific case depends on the environment in which the plastics are placed

The kinetics of plastics degradation depends on whether the environment is dry air, humid air, soil, a landfill, a composting environment, sewage, freshwater, or a marine environment. Each environment has its own characteristic concentration profile of important factors: oxygen, water, other chemicals, daylight, and degrading microorganisms.

According to the nature of the environment, there may be a relatively more efficient, or less efficient, chemical mechanism by which degradation can occur. In one environment a very efficient degradation mechanism might be available, whereas in another environment that mechanism might not be available at all for lack of appropriate conditions. Also according to the nature of the environment, there may be a larger, or smaller, concentration of chemicals that react with the plastic in the degradation process.

For example, is the plastic dry or is it in contact with water? Water plays an important role in speeding degradation by allowing a hydrolysis mechanism to occur. Is the plastic exposed to sunlight or is it buried in the ground or submerged in a body of water, shielded from the rays of the sun? Like exposure to water, exposure to the rays of the sun can strongly affect degradation rates.

The environmental factors affecting the rate of biodegradation include the temperature, moisture level, atmospheric pressure, pressure of oxygen, concentrations of acids and metals, and the degree of exposure to light. Factors relating to the microorganisms include their concentration, whether or not they have enzymes for which the polymer is a substrate, the concentration of enzymes, the presence of trace nutrients for the microorganisms, and the presence of inhibitors or predators.

If any of the required elements is absent, or if it is present at a level that falls below a critical threshhold, biodegradation may not only slow down but may stop altogether until proper conditions are once again present.

(3) Regardless of the environment, the rate of plastics degradation also depends on the chemical composition of the plastic

The degradation rate of a plastic depends on its chemical makeup, and some plastics degrade more rapidly than others because their constituent polymers are chemically different. The rate of biodegradation, in particular, depends on the polymer's characteristics because the polymer is the substrate for the enzyme.

One factor that determines the degradability of a polymer is the nature of the chemical bonds that are present. The chemical structures of the four most common commodity thermoplastics—polyethylene, poly(vinyl chloride), polypropylene, and polystyrene (fig. 4.3)—contain only carbon-carbon single bonds in their backbones; they are polyolefins. That feature makes them particularly resistant to degradation. The resistance of poly(ethylene terephthalate), a polyester, is taken up a bit later.

Poly(vinyl alcohol) (fig. 4.4) is one of the rare polymers having a backbone of carbon-carbon single bonds that is nevertheless biodegradable. The regular occurrence of hydroxyl groups on alternate carbon atoms makes its interactions with water very strong, through the formation of a great many *hydrogen bonds:* an oxygen atom of the alcohol's hydroxyl group is attracted to a hydrogen atom in a water molecule, and an oxygen atom in a water molecule is attracted to a hydrogen atom of the alcohol's hydroxyl group. (Schematically, a hydrogen bond can be represented as O·····H–O.) Poly(vinyl alcohol) is therefore strongly **hydrophilic** (literally, "water loving") and soluble in water, which helps to promote degradation through hydrolysis mechanisms. Poly(vinyl alcohol) is an *exception* to the rule that polymers with carbon-carbon single bonds are resistant to degradation.

Poly(isoprene) is a natural rubber made up of a carbon atom backbone, but the backbone contains carbon-carbon *double* bonds. Natural rubbers are environmentally degradable through the oxidation of the double bonds—by atmospheric oxygen—to produce aldehydes and carboxylic acids. The chains fragment and eventually are reduced in size to the point where biodegradation sets in, followed by complete mineralization. After the natural rubber is *processed* it is very resistant to degradation, not on account of its intrinsic chemical structure, but because of the processing, which includes, for example, the addition of large amounts of antioxidants.

The carbon-carbon single bonds of the polyolefins make them **hydrophobic** (water hating). They are not susceptible to hydrolytic degradation. They can be degraded through oxidative mechanisms but not very readily, and processing increases their resistance. As with rubbers, antioxidants are added to increase their stability.

Besides the nature of the chemical bonds that are present, the details of chain *branching* and even *stereochemistry* (the detailed spatial arrangement of atoms and bonds) are also important. The nature of chain branching and stereochemistry are especially important in biodegradation mechanisms because enzymes are often specific to one particular type of chain branching and one particular stereochemistry. The polymer's molecular weight can also be important, and the degree of chain flexibility.

The *morphology* of the polymer is important as well, including the extent of surface area and degree of crystallinity. The degree of crystallinity is important in the case of polyolefins because oxygen does not easily enter the crystalline regions; they are impermeable to oxygen. Oxidation of polyolefins occurs mainly in the amorphous regions.

The degree of crystallinity is important in the case of polyesters because the crystalline regions are less permeable to water than the amorphous regions, making highly crystalline polyesters particularly resistant to hydrolytic degradation. Poly(ethylene terephthalate) (fig. 4.3), for example, which contains oxygen atoms in its backbone, is nevertheless resistant to degradation on account of its high degree of crystallinity; water has poor access to the hydrolyzable ester group.

Even the degree of crystallinity is partly determined by chemical structure. In poly(ethylene terephthalate) the large intermolecular, chain-chain interactions that lead to high crystallinity are to some extent the result of the strong interactions between the phenyl groups in neighboring chains. Poly(ethylene terephthalate) is so suitable for packaging water and aqueous solutions—like soft drinks and beer—primarily because it can remain in contact with them with no significant hydrolysis taking place.

The degradable synthetic polyesters, poly(glycolic acid) and polycaprolactone (fig. 4.4), are more susceptible to hydrolytic degradation because the ester groups are separated by $-CH_2-$ groups, which impart some flexibility to the chain. Even when they are prepared in highly crystalline forms, as they sometimes are, the flexibility allows penetration of water molecules. In poly(glycolic acid) the ester groups are separated by only one $-CH_2-$ group, and the high density of ester groups along the chain also promotes hydrolysis.

The degradability of a polymer in a plastic also depends on the formulation of the plastic, its processing, and the fabrication of the object. Important factors include the nature of the polymer's interactions with copolymers, with additives, and with coatings. In converting the polymeric material into a plastic, degradability may be substantially reduced by processing and fabrication.

Many synthetic plastics have been manufactured to be resistant to all types of degradation. They take a very long time to degrade after being discarded because they are often totally resistant to biodegradation until sufficient abiotic degradation has occurred. Abiotic degradation is at times made difficult by the chemical nature of the plastics; polyolefins, with their carbon-carbon single bonds, are intrinsically resistant to degradation and stabilizing additives make them more so. Degradation is often further made difficult by their typical environment after disposal.

The complexities of plastics degradation are many, and the goal of producing plastics with controlled lifetimes is, therefore, challenging. Nevertheless, if that goal is pursued with the same vigor that has characterized the pursuit of nondegradable plastics, achieving a wide range of programmed degradable plastics is no doubt possible.

Tests and Standards

It has been recognized for some time that a standardization of *definitions* is necessary in the area of degradable and biodegradable plastics to produce a common language. It is also important to have appropriate experimental *test methods* established for evaluating materials. Ideally the test methods would allow a reliable prediction of the fate and toxicity of plastic materials in any given disposal environment.

Finally, it is important to define what *performance criteria* have to be met in the tests to allow the materials to bear particular labels. Standardization of performance criteria is necessary for several reasons. It is needed by industry managers for product **stewardship**; it is needed for meaningful regulatory programs; and ultimately it is needed for establishing consumer confidence. Confusing product claims often undermine confidence and inhibit product acceptance.

In the United States an acknowledged authority for establishing definitions, test methods, and standards is the American Society for Testing and Materials (ASTM), through its Institute for Standards Research. The European counterpart is the Comité Européen de Normalisation (CEN)—the European Committee for Standardization. Individual European countries have their own organizations, like Germany's Deutsches Institut für Normung (DIN)—the German Institute for Standardization. In Japan the Biodegradable Plastics Society (BPS) has similar responsibilities. The International Standards Organization (ISO) aims to reconcile differences.

ASTM definitions of technical terms used in the plastics industry are the responsibility of ASTM Committee D20 on Plastics, through its Subcommittee D20.92 on Terminology. It originated, for example, ASTM document D 883-00, "Standard Terminology Relating to Plastics." Some of the ASTM definitions from that standard can be found in the Glossary. Several of the most important definitions related to the degradation of plastics are:

degradation, a deleterious change in the chemical structure, physical properties, or appearance of a plastic;
degradable plastic, a plastic designed to undergo a significant change in its chemical structure under specific environmental conditions, resulting in a loss of some properties that may vary as measured by standard test methods appropriate to the plastic and the application in a period of time that determines its classification;
hydrolytically degradable plastic, a degradable plastic in which the degradation results from hydrolysis;
oxidatively degradable plastic, a degradable plastic in which the degradation results from oxidation;
photodegradable plastic, a degradable plastic in which the degradation results from the action of natural daylight;
biodegradable plastic, a degradable plastic in which the degradation results from the action of naturally occurring microorganisms such as bacteria, fungi, and algae;
compostable plastic, a plastic that undergoes degradation by biological processes during composting to yield carbon dioxide, water, inorganic compounds, and biomass at a rate consistent with other known compostable materials and leaves no visually distinguishable or toxic residues.

These definitions have been adopted to be useful in the characterization and evaluation of plastics. A particular definition might be either less inclusive or more inclusive than a definition aimed at serving a more general purpose. The definition of *biodegradable* plastic, for example, refers only to degradation by microorganisms, not macroorganisms. The definition of *compostable* plastic includes a requirement that the residues not be toxic.

Developing test methods for evaluating practical environmental biodegradability is difficult, given the multitude of influencing variables. ASTM's Subcommittee D20.96 on Environmentally Degradable Plastics has responsibilities for biodegradable, photodegradable, and chemically degradable (both hydrolytically and oxidatively degradable) plastics, including test methods, environmental fate, marking, and classification—as well as additional responsibilities for terminology. It originated, for example, ASTM document D 6002-96, "Standard Guide for Assessing the Compostability of Environmentally Degradable Plastics."

That standard guide (D 6002-96) suggests criteria, procedures, and a general approach for establishing the compostability of environmentally degradable plastic materials, including fragmentation rate, biodegradation rate, and safety. The recommended test strategy is three-tiered. Relatively low-cost, rapid-screening tests (Tier 1) aim to determine whether the polymer material is *inherently biodegradable*; that is, biodegradable under controlled, *mesophilic* (20–45 °C), and generally optimized laboratory conditions—which shows that biodegradation *can* occur. Observed biodegradability at that level does not mean the material is *environmentally biodegradable* in the "real world."

A second level includes laboratory and pilot scale assessment (Tier 2). It aims to establish the degradation rate and to confirm biodegradability under *thermophilic* (45–75 °C) laboratory and pilot-scale composting conditions. The results are intended to allow a prediction as to whether the material will biodegrade in the environment. The third level involves full-scale field assessment (Tier 3) based on composting studies with a range of technologies, including backyard composting environments. The assessment is aimed at demonstrating that the plastic material can be expected to biodegrade in the "real world."

Some of the methods used to assess biodegradability include the measurement of carbon dioxide production, as with the **Sturm test** and **soil test**. Other methods involve measurements of **molecular weight** and molecular weight distribution; tensile properties; weight loss; extent of fragmentation; enzyme assays; **biochemical oxygen demand (BOD)**; and ecotoxicity, as with the **cress seed test** and **earthworm test**. Examples of recently developed standards are ASTM Document D 5338-98, "Standard Test Method for

Determining Aerobic Biodegradation of Plastic Materials Under Controlled Composting Conditions," and ASTM Document D 6340-98, "Standard Test Method for Determining Aerobic Biodegradation of Radiolabeled Plastic Materials in an Aqueous or Compost Environment."

Multiple test procedures are necessary in evaluating a material because some tests are subject to false-positive interpretations; that is, concluding incorrectly that degradation or biodegradation has occurred. For example, an observed weight loss may result not from polymer degradation, but from the leaching of additives, including plasticizers. Or, carbon dioxide production might result from the degradation of a low-molecular-weight fraction of the polymer, with no degradation of the longer chains. In another case, a large loss of material strength might come from a very small change in chemical makeup. Strength is often disproportionately affected by the loss of additives, and a 90 percent decrease in strength can result from as little as a 5 percent mineralization.

A distinction is necessary between evaluations of a polymer and a plastic made from that polymer because additives (such as plasticizers and stabilizers), processing, and fabrication may all have effects.

Because of its dependence on many environmental factors, the degradability of a plastic, or polymer, in laboratory evaluations will not be relevant to all disposal environments. Some tests might show it to be potentially or inherently biodegradable without showing it to be actually environmentally biodegradable in a specific disposal environment. Because of concerns over toxicity issues, a distinction is also necessary between an environmentally biodegradable plastic or polymer and an environmentally acceptable biodegradable plastic or polymer; that is, one that leaves no ecologically harmful fragments or residues.

In 1999 ASTM announced its long-awaited "Standard Specification for Compostable Plastics," the ASTM Document D 6400-99. The standard establishes criteria to be met before a product can be labeled compostable. It is based on Standards D 6002-96, D 5338-98, and D 6340-98. A product must, at minimum, satisfy ASTM tests showing conversion to carbon dioxide at 60 percent for a homopolymer or a statistical random copolymer and 90 percent for other types of copolymers and blends in 180 days or less, and leave no more than 10 percent of the original weight on a 2-millimeter screen. If carbon-14 tests are used a test period of 365 days is allowable. Environmental toxicity issues are also addressed, including limits for heavy metals.

Two percentage levels are used (60 percent and 90 percent) because if a homopolymer or statistical random copolymer (i.e., a copolymer with a randomized sequence of the different monomers) degrades at a 60 percent level in 180 days, it can be expected to continue to degrade; the composition of the chain is uniform—or, rather, uniformly random. Nonrandom copolymers include block copolymers, in which the chain is composed of segments of one monomer, separated by segments of another. Sixty percent degradation could merely reflect the degradation of one type of segment, with the other segments being not at all degraded. A 90 percent degradation is needed to indicate that both types of segment are being degraded. Similarly, in blends of two (or more) polymers, a 60 percent degradation could merely indicate that one polymer in the blend has degraded, with the other remaining intact. A 90 percent degradation guarantees that all components of the blend are degrading.

The ASTM Standard is in harmony with German (DIN), European (CEN), and international (ISO) proposals, but there are minor differences. The percent conversion to carbon dioxide in the ASTM specification is to be measured relative to a standard reference material recognized to be biodegradable (e.g., cellulose), whereas the percent conversion in the DIN document is an absolute percentage. The use of a radiolabeled carbon-14 test for 365 days is not present in the CEN document approved in 2000. The ISO will attempt to bring the disparate versions into harmony.

A logo has been proposed that would be used on products that meet the ASTM D 6400-99 specification. The logo was developed jointly by the international Biodegradable Products Institute—a government-industry-academic association that promotes the use of biodegradable polymer materials—and the United States Composting Council—representing the composting industry. Manufacturers can submit their products for independent testing; those that comply with the ASTM specification can bear the logo.

With its relatively short time scale of six months, the ASTM specification is well suited to identifying plastics that will compost quickly, completely, and safely. Composting with total conversion to carbon dioxide in that time frame particularly serves the purpose of waste management; waste is diverted from landfills and incinerators, and litter is reduced.

But composting is now viewed both as a tool of waste management and as a means of promoting agricultural sustainability. For composting to provide mulch and related materials that are long-term soil enhancers, a longer time scale might be suitable. A major value of high-quality compost is in the humic material it contains, in which the carbon atoms have not all been converted to carbon dioxide. Carbon in the plastic could be converted to biomass over a longer time period and converted to carbon dioxide even more slowly, and still be a useful component in compost. The ASTM D 6400-99 specification does not address that issue.

It may be that there are plastics that compost completely and safely, but on a time scale too slow to satisfy the ASTM specification; those plastics still might serve as valuable soil enhancers for agricultural purposes. Indeed, some common, inherently compostable biological materials do not satisfy the ASTM specification, including wood and leaves. It may turn out that ASTM will develop a second standard of compostability aimed at plastics that degrade more slowly. But a longer time scale will make it difficult to collect the experimental data required for establishing an appropriate specification, and the potential impact of undegraded residues on agricultural productivity remains an important issue.

In spite of the technical aspects of defining terms and then testing and assessing plastics, it can still be said that the term *biodegradable* is generally reserved for the ability of a plastic to be assimilated by microorganisms to the point of significant mineralization within a specified time frame and thereby be returned to the ecosystem in a harmless fashion. Moreover, the mineralization of the plastic has to be demonstrated experimentally, with scientific measurements of degradation products such as carbon dioxide.

Inherent or *potential biodegradability* refers to biodegradability under generally favorable conditions of high microorganism concentration and long time periods, indicating a significant potential for ultimate *environmental biodegradability*. For example, polycaprolactone, by its disappearance in a soil burial test method, could be said to be potentially biodegradable in soil. Or, at the other extreme, unmodified polyolefins like polyethylene, polypropylene, and polystyrene can be said to be recalcitrant.

Almost 90 percent of the synthetic polymers now being produced from petrochemicals are inherently nonbiodegradable, and through processing and fabrication they are turned into plastics that are environmentally nonbiodegradable. Their recalcitrance is a fundamental result of their chemical nature. But programmed degradable plastics based on synthetic polymers have been developed and will continue to be developed. Novel catalysts are constantly being discovered, including catalysts that allow an increasing variety of functional groups to be introduced into the polymers. Future research may lead to a larger assortment of programmed degradable synthetic polymers displaying a wide range of degradation behavior.

But now let us expand the horizon in our search for programmed degradable plastics. Instead of restricting ourselves to synthetic polymers made from nonrenewable fossil resources, let us take a look at polymers that occur abundantly in nature, are continually being replenished, and are completely biodegradable. Nature's use of these polymers in living organisms demonstrates their ability to serve as high-performance materials, and illustrates a *functional* similarity between synthetic and natural polymers. Their contrasting *chemical* natures, however, lead to very different degradation behaviors, indicating that naturally occurring polymers can provide renewable feedstocks for biodegradable plastics.

PART TWO
BIOPLASTICS

6

Biopolymers

Nature's Polymers

Plastics manufactured today, with few exceptions, are made from synthetic polymers. But polymers also occur in nature. They are produced by plants, animals, and microorganisms through biochemical reactions. They are *biological polymers,* or simply **biopolymers**.

Carbohydrates and proteins are biopolymers present in the biomass in great abundance; everyone is familiar with them because of their importance in diet and nutrition. Polyesters, produced by microorganisms, are another type of biopolymer; they are less well known but of growing significance. Nucleic acids, which provide genetic material, are biopolymers but much less abundant than the others.

Biopolymers are inherently biodegradable, as they must be in order to take part in nature's cycles of renewal. As we survey the various types of biopolymer, you will see that their *chemical structures are fundamentally different* from those of the synthetic polymers used in almost 90 percent of current plastics production. Biopolymers almost always have oxygen or nitrogen atoms in their polymer backbones; that is the feature that is mainly responsible for their biodegradability. Synthetic polymers that have only carbon-carbon single bonds in their backbones—polyolefins—are particularly resistant to biodegradation.

The biopolymers described here are produced commercially on large scales through mature industries—industries that can provide the feedstocks for new plastics. The large-scale production of these polymers has developed because they have already been found to have important commercial uses, and the new plastics will give added value to these sometimes underutilized components of biomass.

Most production is by extraction from the natural plant or animal material, followed by various steps of purification. Several grades of purity are often processed. For some applications, highly purified polymers are required, as when they are intended for human consumption. Other applications may require less purity, so "practical" grade forms are also produced.

Some biopolymers can be made commercially using microorganisms, like bacteria and fungi, through the process of *fermentation.* The microorganisms convert a supplied substrate into polymers, which are then extracted and purified. Microbial production of biopolymers on very large commercial scales has been developed relatively recently.

There are also polymers that are not found in nature but are made commercially from monomers that are found abundantly in nature. The resulting polymers have the properties of biopolymers, and are referred to here as *honorary biopolymers*.

Carbohydrates

There is more carbohydrate on earth than all other organic material combined. Polysaccharides are the most abundant type of carbohydrate and make up approximately *75 percent of all organic matter.*

The most plentiful polysaccharide is cellulose, found in plant cell walls. It is the most abundant organic compound on earth and alone accounts for *40 percent of all organic matter.* It forms the structural fiber of plants, keeping the cell wall intact and giving it strength. Oven-dried cotton contains about 90 percent cellulose; an average wood has about 50 percent cellulose. Twenty-five to 35 percent of the remaining part of wood is made up of hemicellulose—a generic term referring to polysaccharides that normally occur in plant tissues together with cellulose. A major hemicellulose is xylan. Lignin—described later—makes up 15 to 25 percent of a typical woody plant.

Cellulose cannot be digested by animals. If eaten by ruminants—such as cows or goats—or by certain insects, including termites, it is digested by bacteria that live in the digestive tracts of those animals.

Well over 150 billion pounds of cellulose are produced commercially each year, mainly from wood but also from cotton. Much of it is used in the manufacture of paper and paper products. Cellulose is now also being produced commercially from bacteria by fermentation, but only on a modest scale. Cellulose can be chemically modified in many ways to produce, for example, cellulose acetate. Modified celluloses are used in the manufacture of paint, plaster, adhesives, ceramics, cosmetics, pharmaceutical film coatings, specialty papers, and numerous other products.

Cellulose itself readily biodegrades, beginning with a hydrolysis reaction catalyzed by the extracellular enzyme cellulase. Some types of chemical modification reduce the biodegradability of the polymer.

Starch also is a very abundant component of the planet's biomass. It is found in corn (maize), potatoes, wheat, tapioca (cassava), rice, sorghum, barley, peas, and some other plants. Starch serves an energy storage function in plants. When starch is metabolized by plants, energy is released and that energy is used for plant growth.

The major polymer components of starch are amylose and amylopectin; their relative amounts vary with the source plant. Cornstarch is typically 28 percent amylose and 72 percent amylopectin, but it can be genetically modified to have as much as 85 percent amylose or, for all practical purposes, 100 percent amylopectin (waxy maize starch).

Annual world production of starch is over 70 billion pounds. About 35 billion pounds of starch are produced in the United States each year, mainly by extraction from corn but also from potatoes, wheat, and other sources. The European Economic Community produces nearly 15 billion pounds a year and Japan accounts for 5 billion pounds. Well over half the starch produced is partially hydrolyzed to manufacture corn syrup, dextrose (glucose), and other hydrolysis products that are used as sweeteners or as feedstocks for various manufacturing processes in the chemical, pharmaceutical, and brewing industries. Approximately 75 percent of the remaining starch, either in native form or modified, is used for nonfood industrial purposes in the manufacture of paper and cardboard, paper coatings and sizings, textile and carpet sizing, and adhesives. Food uses include baby foods, bread batter, cake mixes, confectioneries, puddings, pie fillings, glazes, and sauces.

Many common polysaccharides have a basic building block consisting of a six-membered ring of five carbon atoms and one oxygen atom. These ring structures are linked through another oxygen atom. In figure 6.1, (a) shows the chemical structural formula for cellulose; the monomer unit in cellulose is glucose. In the figure, the rings are simplified; everywhere that there are four bonds coming from a single point, the point is meant to represent a carbon atom (C). Two glucose units are shown in the figure in order to illustrate the details of geometry at the linkage.

The structure of xylan is similar to that of cellulose. In xylan the $-CH_2OH$ groups are replaced with H atoms. Hemicelluloses are often variously modified xylans.

The structural formula for one of the major components of starch, amylose, is shown in (b) in figure 6.1. As in cellulose, the monomeric unit in starch is glucose. In the figure, notice the different geometrical arrangement of atoms at the linkage of amylose compared with that in cellulose. The position of the linkage oxygen atom (O) and that of the hydrogen atom (H) attached to the same carbon atom are interchanged in the two structures. That is the only difference in the chemical structures of cellulose and amylose. In amylopectin there are branch points in the chain where segments of chain, identical in chemical structure to the main chain, are attached to the main chain where the $-CH_2OH$ group is located.

The structure of chitin, (c) in figure 6.1, is identical to that of cellulose except that a hydroxyl group (–OH) on every ring is replaced with an acetamido group ($-NHCOCH_3$). In chitosan the $-COCH_3$ group of chitin is replaced with H.

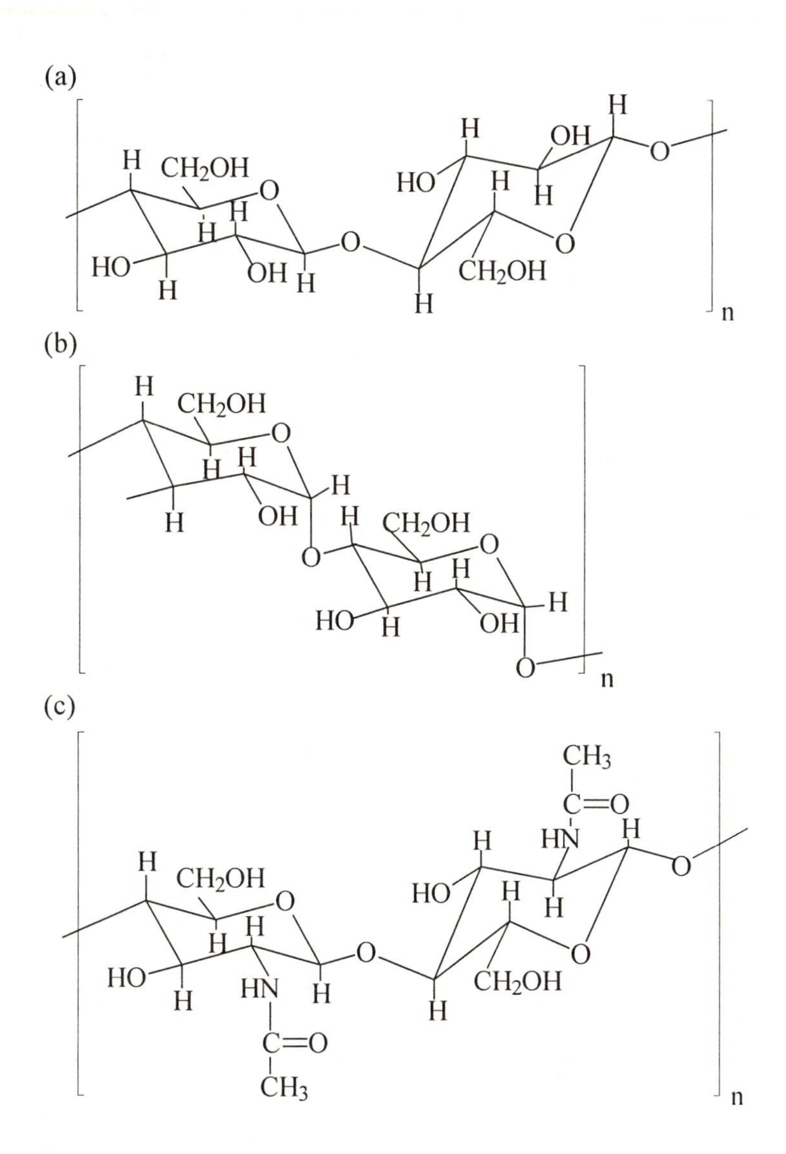

Figure 6.1 *Structural formulas of (a) cellulose, (b) starch, and (c) chitin*

Chitin, another abundant polysaccharide, is found in the skeletal tissue of marine crustaceans—shrimp, crab, lobster, and other shellfish; in the skeletal tissue of insects; and in the cell walls of many filamentous fungi. In crustaceans and insects, chitin is one part of the hard composite material forming the exoskeleton.

Chitin is produced mainly from shellfish waste. An important derivative, chitosan, is produced commercially from chitin by a base-catalyzed deacetylation reaction. The world market for chitin and chitosan is only 2 to 4 million pounds a year, a very small amount compared with its natural abundance. Both chitin and chitosan have applications in cosmetics and personal hygiene products, in agriculture, and as food ingredients. There is strong interest in finding additional applications for this ample biopolymer.

Large amounts of polysaccharides are also present in seaweeds. Agar and carrageenan are found in various red seaweeds; alginates are common to brown seaweeds. Alginate is also produced by bacteria.

The major component of agar is agarose, whose chemical structure is shown in (a) in figure 6.2. Carrageenan is shown in (b) in figure 6.2. Notice that agarose and carrageenan have very similar chemical structures.

Agar is widely used as a culture medium in microbiology. It is also an additive for confectioneries, desserts, beverages, ice cream, pet foods, and health foods. Nonfood applications include toothpaste, cosmetics, medical ointments, adhesives, corrosion inhibitors, and photographic emulsifiers. The world market for agar is around 15 million pounds a year.

Carrageenan is used as an emulsifier in food products, as a stabilizing aid in ice cream and process cheese, and in toothpastes, cosmetics, and pharmaceuticals. Its world market is 30 to 35 million pounds a year. Alginate, with a world market of 30 to 40 million pounds a year, has various commercial uses as a thickener, with such diverse applications as a dye thickener in textile printing and as a thickener in salad dressings.

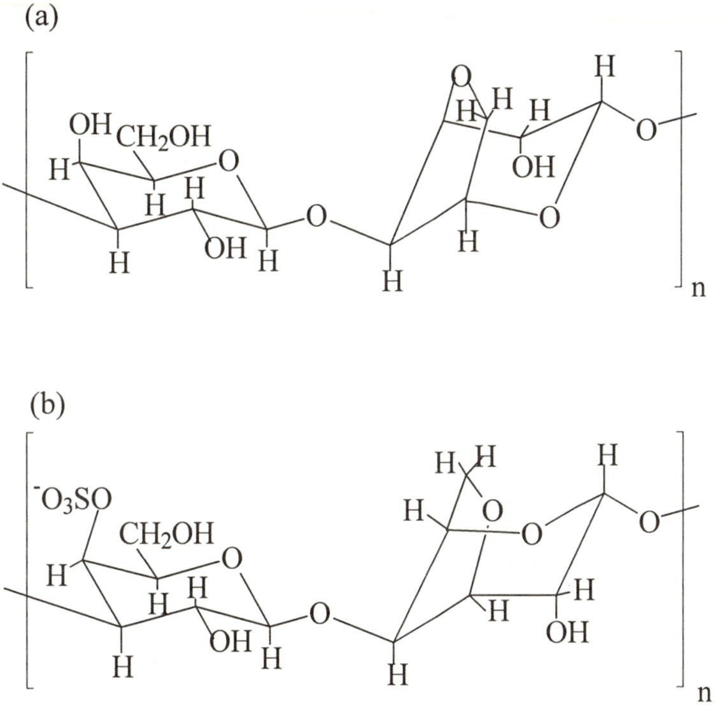

(a)

(b)

Figure 6.2 *Structural formulas of (a) agarose and (b) carrageenan*

Agarose, carrageenan, and alginates are gelling polysaccharides; all have been used to form encapsulating gel beads for drug delivery systems.

Other commercially important plant polysaccharides are pectin and the guar, locust bean (carob), tara, and fenugreek gums. They are common food additives and are variously used as thickening agents and stabilizers in jams and jellies, in canned fruits, and in dairy products—especially ice cream and cheese products. Some are used in pharmaceuticals, paper coatings, textile printing, and water-base paints.

Higher animals contain glycosaminoglycans, a class of complex polysaccharide that includes hyaluronan. Hyaluronan, which is commercially extracted mainly from rooster combs, has a variety of biomedical applications, including use in eye and ear surgery, wound treatment, and arthritis treatment. Other glycosaminoglycans are heparin, dermatan, and chondroitin.

Bacterial and fungal polysaccharides prepared through fermentation have become commercially signficant more recently. Bacterial polysaccharides include xanthan, dextran, gellan, welan, rhamsan, curdlan, polygalactosamine, and levan. Xanthan is produced in the largest amounts, currently 20 to 40 million pounds a year. Large-scale fermentation is used in its production, with molasses or corn syrup as feedstock, and the aerobic bacterium *Xanthamonas campestris* as the biological agent. Xanthan is widely used as a thickener in food applications, as a stabilizer in process cheese, and in miscellaneous industrial applications. Dextran has medical applications as a blood plasma extender, in wound coverings, and in other applications. Gellan is used in the food industry as a gelling agent. An example of fungal polysaccharide is pullulan, found in yeast. Pullulan is used as a food additive to enhance bulk and texture; as a food coating; in medicine as a plasma extender and in adhesives; and in other miscellaneous applications.

The abundance of polysaccharides makes them prime candidates for becoming major *renewable* feedstocks for new plastics.

Lignin

Lignin (from the Latin *lignum,* meaning wood) is the second most abundant component of woody material; it accounts for about *20 percent of all organic matter.* Lignin is a largely amorphous material of complex structure and is currently pictured as a cross-linked network macromolecule. It is isolated in large amounts in the processing of wood pulp—about 100 billion pounds a year in the United States—but most of it is simply burned as a fuel. The small remainder—around 90 million pounds (or 0.1 percent)—has various industrial uses in concrete admixtures, animal feed binder, gypsum wallboard, adhesives, and others. There is great interest in finding additional applications for this plentiful but underused resource. The overall economics of extracting the carbohydrates (cellulose and hemicelluloses) from woody material would be improved if more uses for lignin could be found; in addition to the value of the new uses, its disposal costs could be avoided.

Attempts to develop extensive markets for lignin have so far failed. New technologies for pretreating lignocellulosic biomass are being developed on laboratory scales, and the most promising of them may prove to be ultimately commercially competitive for producing large amounts of industrial feedstocks from woody plants.

Proteins

Proteins are polymers formed by the condensation polymerization of amino acids. The carboxyl group of one amino acid joins with the amino group of another amino acid, eliminating a water molecule and forming an amide linkage, called in the case of proteins a peptide bond. Each specific protein is made up of a characteristic sequence of amino acids.

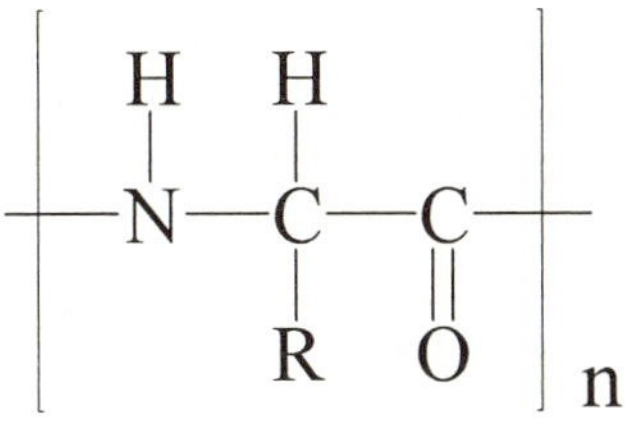

$$\left[-\mathrm{N(H)}-\mathrm{C(H)(R)}-\mathrm{C(=O)}- \right]_n$$

Figure 6.3 *Structural formula of a protein*

The structural formula for a generic protein, in terms of its amino acid residues, is shown in figure 6.3, in which R varies from residue to residue and identifies the amino acid from which each residue is derived. If the complete protein sequence is known, the chemical formula can be expressed by naming each residue in order.

There are twenty amino acids that are common to all living organisms: glycine (gly), alanine (ala), phenylalanine (phe), valine (val), leucine (leu), isoleucine (ile), aspartic acid (asp), glutamic acid (glu), tyrosine (tyr), asparagine (asn), glutamine (gln), lysine (lys), arginine (arg), serine (ser), threonine (thr), cysteine (cys), methionine (met), histidine (his), tryptophan (trp), and proline (pro).

Some proteins have a structural function—like collagen and keratin; others are enzymes—the catalytic agents of biochemical reactions. The function of a protein is reflected in and originates in its three levels of structural organization. The amino acid sequence of a protein constitutes its *primary structure.* Some individual segments of the sequence are often predisposed to form regular, helical conformations, and the array of ordered helical segments defines the protein's *secondary structure.* Structural proteins have long segments of chain in ordered helical geometries, a property which leads to the formation of fibrous aggregates. Enzymes, on the other hand, often have a series of shorter ordered segments separated by amorphous regions, a feature which allows the protein chain to fold into a compact globular conformation unique to that enzyme, giving rise to a third level of organization called its *tertiary structure.*

The chain associations in structural proteins and the folded tertiary structures in enzymes are stabilized by a combination of physical interactions—like those that lead to hydrogen bonds—and full-blown chemical bonds—such as disulfide bonds. (The side chain of cysteine [$–CH_2–SH$] contains the sulfhydryl [–SH] group. The oxidation of two cysteines, in proximity to one another, produces a disulfide bond $–CH_2–S–S–CH_2–$.) Disulfide bonds provide important chemical cross-links in protein structures.

Among the abundant animal proteins are collagen, casein, whey protein, and keratin. Collagen is the most abundant protein found in mammals and typically constitutes 30 percent of total protein. It is an insoluble fibrous protein found in many animal connective tissues, and it is present in animal skins and hides, blood vessels, tendons, ligaments, and ossein, the connective-tissue protein of bones.

Each polypeptide chain of collagen consists of a repeating tripeptide sequence containing the common residue glycine, as in gly–X–Y, where X is often the common residue proline and Y is often the uncommon residue hydroxyproline. In collagen, long chains of the polypeptide form triply stranded helices. These triple helices arrange themselves in groups to form fibrils, which give strength to bones and allow them to flex under stress.

Gelatin is denatured collagen in which the triple helices of collagen are disrupted and partially hydrolyzed, leaving polymer chains that are largely disordered and unorganized. Gelatin has a long history of commercial use. In the United States it is mainly prepared from hog and cattle skins; in Europe it is often extracted from animal bones. Large amounts of gelatin are used as a thickener for desserts, ice cream, confectioneries, and baked goods. It is also used to make sausage casings and to form capsules for drugs and vitamin preparations. As a hot melt it has excellent adhesive properties and is widely used in bookmaking. It has other miscellaneous industrial applications, including use as a stabilizer-binder in photographic emulsions.

Casein is found in milk. The sequences of some fragments of casein are known; one fragment contains the eighteen amino acid sequence: asn-ser-ile-asn-thr-gly-glu-glu-ser-glu-glu-ala-ser-ser-ser-ser-ile-ile. Casein, predominantly from skimmed cow's milk, is used as a labeling adhesive in the bottling industries on account of its excellent rheological properties, and in binders, protective coatings, leather finishes, and other products.

Whey is the soluble fraction of milk that is separated from the casein curd during cheese manufacture. Whey protein, isolated from whey, has several important applications in the food industry, including uses as a body and bulking agent.

Plant proteins are also important commercially, including soy protein, zein from corn (maize), wheat gluten, potato proteins, and pea proteins. Soybeans consist of approximately 30 to 45 percent protein and 20 percent oil; the remainder is mostly carbohydrate material. Soy protein is used for making plywood adhesive and coatings for paper and paperboard.

In corn, approximately 40 percent of the kernel protein is zein, constituting about 4 percent of the kernel weight. Zein is used as a binder in printing inks, as a shellac substitute, in floor coatings, and in coatings for grease-proof paper—on account of its high barrier toward oils and fats. Zein can also be formed into fibers and films that are tough, glossy, and scuff resistant. It is water insoluble and thermoplastic.

Commercial proteins—as a group—are generally more expensive than polysaccharides, but new uses would give them added value, perhaps providing an incentive for increased production. They are a potentially large source of feedstocks for biodegradable plastics.

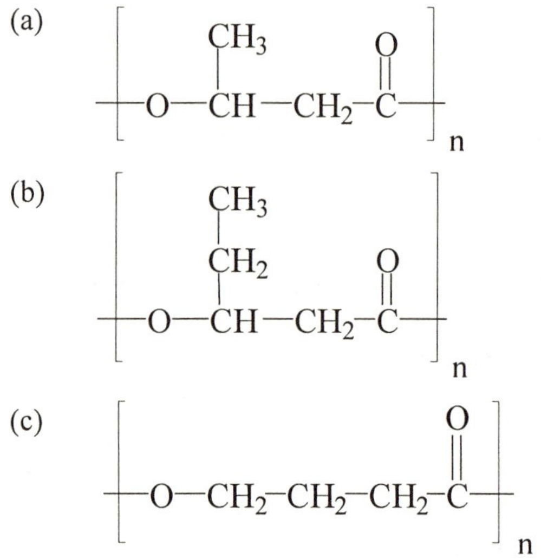

(a) $$\left[\!-\!O\!-\!CH(CH_3)\!-\!CH_2\!-\!C(=O)\!-\!\right]_n$$

(b) $$\left[\!-\!O\!-\!CH(CH_2CH_3)\!-\!CH_2\!-\!C(=O)\!-\!\right]_n$$

(c) $$\left[\!-\!O\!-\!CH_2\!-\!CH_2\!-\!CH_2\!-\!C(=O)\!-\!\right]_n$$

Figure 6.4 *Structural formulas of some bacterial polyhydroxyalkanoate polyesters: (a) poly-3-hydroxybutyrate (PHB), (b) poly-3-hydroxyvalerate (PHV), (c) poly-4-hydroxybutyrate*

Polyesters

Bacteria produce polyesters, including polyhydroxyalkanoates (pronounced póly-hydroxy-álkan-oh-ates)(PHAs), which typically occur as inclusion bodies deposited in granules in the cytoplasm. PHAs serve as energy and carbon storage materials in bacteria. PHAs accumulate when carbon is in excess but some other nutrient limits growth. They are consumed when no external carbon source is available. The most abundant PHA in nature, and the first to be discovered, is poly-3-hydroxybutyrate (PHB), shown in (a) of figure 6.4. Other PHAs have since been discovered, including those containing hydroxyvalerate units, as in poly-3-hydroxyvalerate (PHV), shown in (b) of figure 6.4. In figure 6.4, (c) shows a structural variant of PHB.

Naturally occurring polyesters, like the polyhydroxyalkanoates shown in figure 6.4, are now produced commercially from microorganisms through fermentation. In *bioreactors,* microorganisms are fed a carbon source substrate, such as glucose or sucrose for

PHB, or propionic acid for PHV. Propionic acid can be produced by the fermentation of wood pulp waste or from petroleum.

Polyhydroxyalkanoates are produced in a two-stage fermentation process consisting of cell growth followed by polymer accumulation, which proceeds to as much as 80 or 90 percent of the cell's dry weight, as is the case with *Ralstonia eutropha* (formerly named *Alcaligenes eutrophus*). To harvest the polymers the cell wall is ruptured and the polymers are collected and purified.

Production of polyhydroxyalkanoates has reached approximately a million pounds a year. They are biocompatible as well as biodegradable and have biomedical applications in the areas of controlled drug release, surgical sutures, and others. These natural polyesters are another potentially large feedstock for biodegradable plastics.

Carbohydrates (polysaccharides), proteins, and polyesters together make up a large and chemically diverse supply of renewable polymers, ready for use in the production of biodegradable plastics (Chapter 7).

Synthetic "Biopolymers"

Many monomers that occur naturally in the biomass can be polymerized. The resulting polymers might be called *honorary biopolymers*, because while the polymers themselves are not produced by living organisms, they have the properties of biopolymers and their monomers originate in living organisms. Most important, they are biodegradable. Lactic acid, various amino acids, and various triacylglycerols are examples of important polymerizable biomolecules.

Lactic Acid and Poly(lactic acid)

Lactic acid, $CH_3CHOHCOOH$, occurs naturally in animals and in microorganisms. It is found in many natural foods and in fermented foods such as yogurt, buttermilk, sourdough breads, sauerkraut, and others. It is widely used commercially in food-related applications as a preservative and as an emulsifying agent in bakery products.

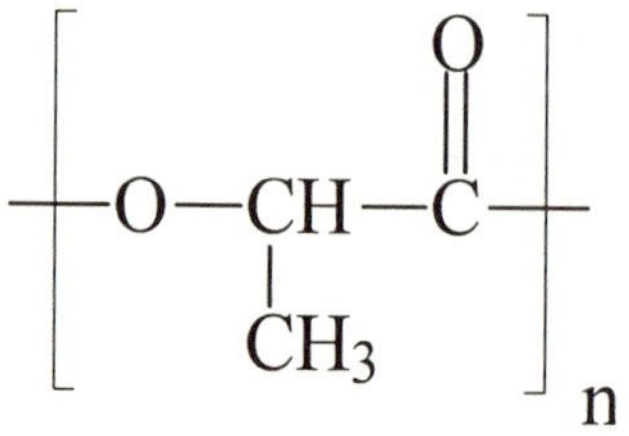

$$\left[-O-\underset{\displaystyle CH_3}{\underset{|}{CH}}-\overset{\displaystyle O}{\overset{\|}{C}}- \right]_n$$

Figure 6.5 *Structural formula of poly(lactic acid), a synthetic "biopolymer"*

Lactic acid can be produced commercially by chemical synthesis or by fermentation; fermentation is now the major route. In the fermentation process sugar feedstocks, such as dextrose (glucose), are obtained either directly from sources such as sugar beets or sugar cane, or through the conversion of starch from corn, corn steep liquor, potato peels, wheat, rice, or some other starch source.

Yields of lactic acid are greater than 90 percent, which helps make the fermentation step inexpensive. In batch production lactic acid can be produced at the rate of 2 grams per liter per hour. The lactic acid is recovered from the fermentation broth and purified in a multistep process, which represents a major part of the production costs.

Poly(lactic acid) (PLA) is the polyester synthesized from lactic acid; its structure is shown in figure 6.5. It is also called polylactate or polylactide.

Condensation polymerization of lactic acid generally produces a low-molecular-weight polymer, which is then treated with coupling reagents that act as chain extenders to give high-molecular-weight PLA. Recently a condensation polymerization has been found to give high-molecular-weight PLA when carried out in a high-boiling-point solvent at reduced pressure and with added catalyst.

Alternatively, the low-molecular-weight *prepolymer* is depolymerized under reduced pressure to produce lactide, the cyclic dimer of lactic acid (a dilactone). Metal-catalyzed ring-opening polymerization of lactide produces a high-molecular-weight polymer with good mechanical properties. The name polylactide refers in particular to the product of ring-opening polymerization.

PLA has various industrial applications, including use in water treatment as a flocculent—an agent that serves as a site for the aggregation of fine suspended matter, enhancing the gravitational settling rate and removal of particulate impurities. PLA has recently become very important commercially as a leading bioplastic material (Chapter 7).

Amino Acids and Poly(amino acids)

Amino acids can also be polymerized. One example is poly-(aspartic acid), which has been described as a biodegradable substitute for synthetic polyacrylate, for use as a mineral scale inhibitor in water-treatment applications, a dispersing agent in detergents to prevent the deposition of soil on surfaces, and a dispersing agent for pigments in paints.

Triacylglycerols (Triglycerides) and Their Polymers

An abundant class of biomolecule that has only recently received attention as a candidate feedstock for polymer resins are the triacylglycerols, commonly called triglycerides. They are the most abundant class of compound in the family of compounds known as *lipids.* Triglycerides make up a large part of the storage lipids in animal and plant cells. When liquid at room temperature they are called *oils,* and when solid they are *fats*—a somewhat imprecise classification since only a few degrees' change in temperature will cause many of them to change from solid to liquid or liquid to solid.

Commercially important oils are produced from the seeds of soybeans, corn (maize), cotton, sunflowers, flax (linseed), rapeseed, castor beans, tung, palms, peanuts, olives, almonds, coconuts, and canola. Flax (linseed) and tung oils are examples of *drying oils*—those that form a tough, elastic film on drying and are used in paints, varnishes, and enamels. Soybean and corn (maize) oils are examples of *semidrying oils,* used for cooking and in foods. Castor and rapeseed oils are examples of *nondrying oils,* which remain greasy and eventually turn rancid. Canola is the result of a breeding program undertaken to remove undesirable traits from rapeseed. Over 16 billion pounds of vegetable oils are produced in the United States each year, mainly from soybean, flax, and rapeseed.

Lipids are water-insoluble biomolecules whose structure is mainly hydrocarbon in nature. *Fatty acids* are lipids made up of long hydrocarbon chains, saturated or unsaturated, and a terminal carboxyl group. In figure 6.6, (a), (b), and (c) show the chemical structural formulas of representative fatty acids: palmitic acid, oleic acid, and linoleic acid. When all three hydroxyl groups of the alcohol *glycerol*, (d) in figure 6.6, are esterified with fatty acids, the structure is a *triacylglycerol*, shown as (e) in figure 6.6. R, R', and R" stand for the hydrocarbon chains of the particular fatty acid at each of the three positions.

The fatty acid composition of vegetable and animal oils varies with the source and largely determines their use. Soy oil, for example, contains about 55 percent linoleic acid, 22 percent oleic acid, and 10 percent palmitic acid (fig. 6.6). It is well suited for edible oil applications. Soy oil accounts for 80 percent of the seed oils produced in the United States and 30 percent of the world's supply of vegetable oil. Approximately 600 million pounds a year are used in the United States in nonfood applications.

Worldwide, three-quarters of seed oil production is for edible oil, but the large remainder is used in industrial applications such as surfactants, coatings, lubricants, adhesives, drying agents, cosmetics, printing inks, emulsifiers, and plasticizers.

Plant oils that are mainly used in industrial nonfood applications often have unusual chemical compositions. Castor oil, one of the most important industrial plant oils, contains ricinoleic acid, which is identical to oleic acid, shown as (b) in figure 6.6, except that carbon-12 (counting the –C=O carbon as carbon-1) has a hydroxyl (–OH) group in place of one of the hydrogen atoms. That single change increases the viscosity of ricinoleic acid relative to oleic acid, and enhances its value for industrial applications. It is

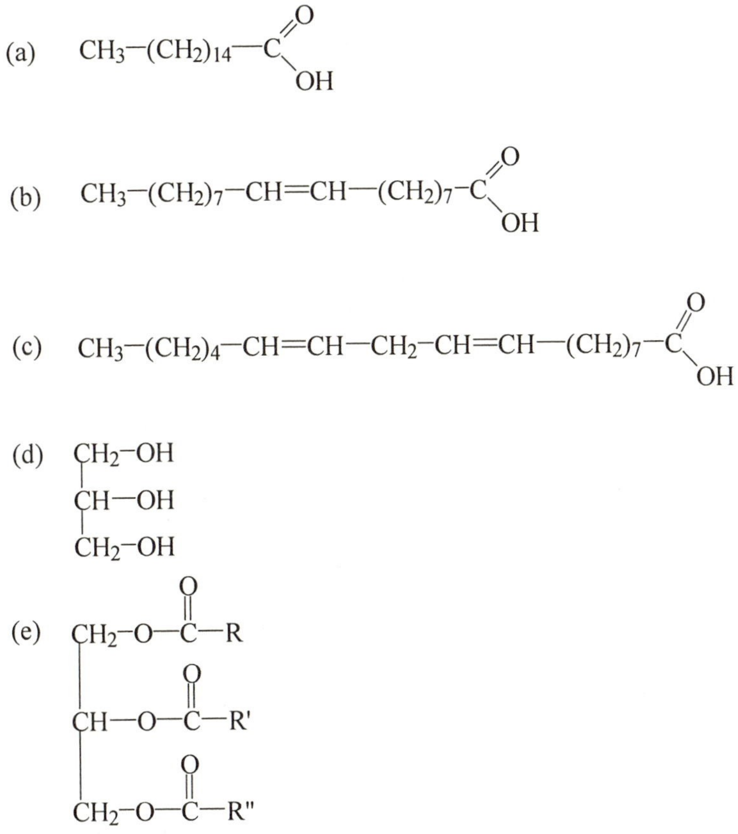

(a) $$CH_3-(CH_2)_{14}-C(=O)-OH$$

(b) $$CH_3-(CH_2)_7-CH=CH-(CH_2)_7-C(=O)-OH$$

(c) $$CH_3-(CH_2)_4-CH=CH-CH_2-CH=CH-(CH_2)_7-C(=O)-OH$$

(d) $$\begin{array}{l} CH_2-OH \\ | \\ CH-OH \\ | \\ CH_2-OH \end{array}$$

(e) $$\begin{array}{l} CH_2-O-C(=O)-R \\ | \\ CH-O-C(=O)-R' \\ | \\ CH_2-O-C(=O)-R'' \end{array}$$

Figure 6.6 *Structural formulas of (a) palmitic acid, (b) oleic acid, (c) linoleic acid, (d) glycerol, and (e) a triacylglycerol*

used in lubricating oils and greases, coatings, sealants, and as a plasticizer for nitrocellulose. The additional –OH group also allows polymer formation, and ricinoleic acid is used in plastics and rubbers.

Castor seeds contain over 50 percent oil on a dry-weight basis, of which nearly 90 percent is ricinoleic acid. Castor beans used to be a major crop in the United States; production peaked in the 1950s and 1960s. It was abandoned in 1972 because of a dispute between seed processors and oil buyers. Now castor beans are grown mainly in India and Brazil. The United States imports $30 million worth of castor oil annually.

The polymerization of triglycerides has recently received renewed attention in the context of producing new bioplastic materials (Chapter 7).

Nature's Fibers

Natural fibers, like cotton, wool, and silk, illustrate nature's use of biopolymers to form strong materials. Because of their strength they have been used for textiles for thousands of years. Wool is known to have been used 35,000 years ago, cotton for 14,000 years, and silk for at least 5,000 years.

Cotton is a plant fiber made up of approximately 90 percent cellulose. Wool is the cut hair of sheep, or sometimes of goats or other animals. Around 80 percent of the composition of wool is the protein *keratin* and 17 percent is other proteins; the remaining 3 percent is made up of polysaccharides and other molecules.

Silks are externally spun fibrous protein secretions; they are the only natural fibers that are spun by the producing organism. The word *silk* usually refers to the fiber that is mainly produced from cocoons of the mulberry silk spinner (silkworm), although a similar natural fiber is made by spiders and other insects. The major biopolymer component in silk, approximately 78 percent of it, is the protein *fibroin*; the remaining 22 percent is *sericin,* an adhesive protein that is sometimes called silk glue. (The domestic use of silkworm silk is *sericulture*.)

Silk's combination of strength and flexibility is the direct result of the polymer's chemical structure. Within the fibroin molecule there are regions of approximately sixty amino acid residues consisting of a repeated hexapeptide sequence, ser-gly-ala-gly-ala-gly. These sixty residue regions form ordered helical chain conformations, but are separated within the chain by more flexible units of around thirty amino acid residues. The ordered segments pack together in bundles to give silk its strength; the amorphous connecting segments provide reduced crystallinity and flexibility.

Industrial fibers—those used for nontextile purposes—include flax, hemp, sisal, ramie, jute, and bagasse (the residue from sugar cane); they all contain cellulose. They are used for rope and cord, agricultural bindings, filter cloths, and reinforcing fillers for plastics.

Natural fibers can be *very* strong. Silkworm silk is as strong as nylon; spider silk is stronger than nylon. On a strength-to-weight basis both outperform steel. The tendon-like fibers used by marine mussels to latch so strongly onto ships and rocks are made of protein material, including collagen and elastin.

Natural fibers illustrate the *high-performance* character of natural biodegradable materials. But even more impressive is the way living systems have evolved to create high-performance and programmed degradable composites.

Nature's Composites

Through evolution some extraordinary materials have been produced, serving complex and diverse functions. Strength is sometimes combined with elasticity, as in blood vessels and skin; sometimes strength is combined with less elasticity but still with the ability to flex under stress, as in bones; and sometimes strength is combined with extreme hardness, as in teeth. In its design of materials, nature uses a heirarchical system of structure combined with the use of composites.

Both principles are well illustrated in the external structural tissues of animals. Skin, hair, fingernails, fur, feathers, beaks, claws, and horn all contain the protein *keratin.* In these tissues helical keratin molecules form an intertwined right-handed double helix, and two double helices twist about each other in the opposite left-handed sense, to form a *protofibril*. (The same strengthening principle is used by engineers in the coiled steel cables of suspension bridges!) Groups of eight protofibrils pack together to form microfibrils, which are the basic structural building blocks in keratin-containing tissue. The heirarchical strategy of forming molecules into double helices, then protofibrils and microfibrils, provides for a strong fundamental building component.

But then the microfibrils are cross-linked with disulfide bonds to varying extents, depending on the tissue. A low level of cross-linking in skin allows for flexibility and softness. Greater cross-linking produces the hardness of claw and horn. The cross-linked keratin fibrils, finally, are embedded in a matrix of other proteins to produce a *fiber-reinforced composite*. An important matrix protein is *elastin*, which, as its name suggests, is mainly responsible for the elastic properties of such tissues as skin, blood vessels, ligaments, and lungs.

Collagen, already described, occurs in many different internal structural tissues including blood vessels, ligaments, tendons, and teeth. Collagen, like keratin, is arranged in multiple-chain fibrils, again illustrating the heirarchical structural principle. It too is strengthened with cross-links. In collagen the cross-links are forged between hydroxylysine residues. Hydroxylysine is an unusual amino acid, but in collagen it is present to the extent of about 1 percent, a small variation in the overall chemical makeup, but enough to produce remarkable effects.

The extent of cross-linking varies from one tissue type to another. In blood vessels, softness and elasticity are required. Those properties are provided by a low cross-linking level and the presence of elastin. With more cross-linking, collagen forms strong, but still elastic, connections between bones (ligaments) and attachments of muscles to bones (tendons). A matrix of other proteins completes the composites. In teeth, collagen is combined with a hard inorganic crystalline material to form yet another fiber-reinforced composite.

Nature's design for plants is also impressive. Wood, as a material, has to be strong enough to support the numerous green leaves required for photosynthesis. But wood must also provide circulatory systems for the transport, throughout the tree, of water and of the sugars produced by photosynthesis. Strength is provided by cellulose-fiber cell walls. Within the cell walls the cellulose fibers are laid out in criss-cross reinforcing patterns and are embedded in a matrix of other molecules, mainly lignin. In the space between cells are other organic molecules, including hemicellulose and (more) lignin.

Strength and some degree of elasticity are only two of the required properties in structural tissues. Transport of molecules through tissues also has to be controlled. Skin, for example, has to inhibit the passage of infectious agents, like bacteria, into the body. At the same time it has to allow moisture transport outward, as perspiration, but not so much as to dry out the body or skin; water is an important plasticizer for skin. Skin, with its strength, elasticity, and transport properties, makes an excellent *packaging material*.

Helping to control the transport of water are oils and waxes. They are found as protective coatings on skins, furs, and feathers of animals, on the leaves and fruits of higher plants, and on the exoskeleton of many insects.

Not only do nature's composites provide the required strength, elasticity, barrier properties—and other properties, but they also biodegrade in a programmed manner. Very rapid environmental biodegradation is generally not the rule. When a tree dies, the wood biodegrades in stages: the hemicellulose degrades most rapidly, the cellulose next, and the lignin degrades at a slower rate, providing an important soil stabilizer that maintains organic content and moisture level. The mix of multiple time scales helps promote a proper ecological balance.

If we need examples of high-performance materials that are also programmed degradable, we only have to look to nature!

7

The Reemergence of Bioplastics

What Are Bioplastics?

The natural abundance of biopolymers, the variety of their chemical structures, their nontoxicity, and their biodegradability combine to make biopolymers a large potential source of feedstocks for biodegradable plastics. **Bioplastics** is a concise—and suitable—name for *biodegradable plastics whose components are derived entirely or almost entirely from renewable raw materials.* A *bio*plastic contains one or more *bio*polymeric substances as an essential ingredient. They are indeed plastics, but very special plastics.

Although biopolymers are inherently biodegradable, processed materials may or may not be a substrate for the same enzyme system that degrades the biopolymer. For a bioplastic to be practically biodegradable in a given environment requires that it be a substrate for microorganisms found in that environment. Bioplastics have to pass the test that they are practically, or environmentally, biodegradable.

In nature, biopolymers are plasticized by components in their immediate surroundings. During the commercial processes of extraction and purification, the native plasticizing agents may be removed, but plasticizing agents can be reintroduced to produce nonbrittle bioplastics.

Other additives can enhance their properties. The final formulation might consist of one or more biopolymers combined with one or more plasticizing agents, and one or more other additives. Each component will contribute particular properties to the final composition.

biopolymer(s) + plasticizer(s) + other additive(s)

= BIOPLASTIC

Some biopolymers are thermoplastic and can be processed with the same methods used with synthetic polymers, such as extrusion or compression molding. Products can be processed into three dimensional objects as well as flat sheets. With biopolymers that are not thermoplastic, sheets can be made by casting; as solvent is removed a gel state is typically produced and continued solvent removal results in solid bioplastic material. Bioplastics technologies are often water-based because many biopolymers are soluble or dispersable in water, or can be made to be so. Some naturally occurring low-molecular-weight biomolecules can be polymerized, at times forming thermoplastic materials. In other cases the final stage of polymerization might involve cross-linking to produce strong thermosets.

Early Bioplastics

Bioplastics are not new. In the biblical Book of Exodus, Moses' mother built his ark from rushes, pitch, and slime, a composite that might now be called a fiber-reinforced bioplastic. Natural resins—like amber, shellac, and gutta percha—have been mentioned throughout history, including during Roman times and the Middle Ages.

Animal bones—based on the protein collagen—and horns and hooves—containing the protein keratin—were sometimes boiled in water or oil, heated directly, or soaked in alkaline solution, then molded or pressed into sheets. Horn was used, perhaps as early as the thirteenth century, to make drinking beakers and other containers. In North America, Native Americans were developing and refining techniques for making horn ladles and spoons long before there was any European contact. The bowl section was thinned,

Figure 7.1 *Native American horn spoon (11½" x 3"). Horn spoons and ladles date back to ancient times; they were used as feast utensils and often were elegantly carved. One-piece spoons were the original form. The two-piece spoon shown here (ca. 1860) represents a later technological advance in the use of materials. The mountain goat horn handle retains the natural curvature of the goat horn; the bowl was formed by steaming and shaping a sheep horn. The bowl is joined to the handle with copper rivets. (The Seattle Art Museum, Gift of John H. Hauberg. Photograph by Paul Macopia)*

made pliant, then shaped to dimensions wider than the original horn, while the handle retained the natural curvature of the horn's spiral shape (fig. 7.1).

Molded horn snuff boxes were being manufactured by the eighteenth century, and horn continued to be popular throughout the nineteenth century. In the National Museum of American History's collection, there is a lantern made around 1812 that has translucent horn windows (fig. 7.2).

Figure 7.2 *Tin-plated sheet-iron lantern (18.9") with translucent horn windows (ca. 1812). Manufacturing horn items was a significant industry through the eighteenth and nineteenth centuries. Horn items included combs, jewelry, snuff boxes, shoe "horns," and many other items. (National Museum of American History, Smithsonian Institution, Washington, D.C.)*

Significant commercialization of bioplastics began to be developed in the nineteenth century. By the middle of the century, ebonite—a black vulcanized form of natural rubber—was being processed into articles like combs, buttons, brooches, and electrical insulation. Gutta percha—like rubber, an extract from tropical trees—was being made into undersea cable insulation, water hoses, ornamental frames, and other objects. A screw extruder was developed for its processing, an early precursor of modern extrusion machines. At midcentury, 8 percent of all patents filed in Great Britain involved plastics.

In 1845 Christian Friedrich Schönbein, a German-born chemistry professor, was teaching at the University of Basel, Switzerland. Following earlier work by two French chemists, Théophile-Jules Pelouze and Henri Braconnet, he prepared a cellulose derivative from paper (cellulose nitrate) that was strong, transparent, waterproof, and, as he wrote to Michael Faraday, "capable of being shaped out into all sorts of things and forms." It also attracted attention as an explosive.

Louis Ménard, a student of Pelouze, discovered in 1846 that a solution of cellulose nitrate in ethanol, called *collodion,* dried into a tough, elastic, and waterproof solid but found no practical application for it. In 1848, J. Parker Maynard of Massachusetts discovered that collodion, when spread on open wounds, dried to form an airtight, watertight dressing that promoted healing. And in 1851, Frederick Scott Archer, an English sculptor, used collodion as the basis for a duplicative technique in producing photographs.

The English inventor Alexander Parkes produced a pressure-molded product from collodion that he called *Parkesine.* At the 1862 International Exhibition in London he displayed many heat-softened and pressure-molded objects, including boxes, artificial teeth, combs, buttons, knife handles, and other objects, some of which are held today in the London Science Museum's collection. He received a bronze medal for excellence of quality and ingenuity (fig. 7.3).

In 1865, Parkes reported the improvement in texture uniformity and contractile properties brought about by adding 2 to 20 percent camphor to his formulation, and in 1866 he launched a commercial venture to market Parkesine. Commercially produced items included umbrella handles, combs, chess pieces, earrings and bracelets, and even a few billiard balls and dentures.

But after only two years, Parkes's company failed. The failure has been attributed to a combination of factors, including the flammability of Parkesine, over-rapid production, and the use of inferior raw materials arising from pressure to keep costs to the consumer down. At times cotton rags provided the source of cellulose, instead of the cotton linters used in the early high-quality exhibition articles (fig. 7.3). Patent rights were reassigned as part of the liquidation, and the assignee, Daniel Spill—a business associate of Parkes's—continued commercialization through the Xy-

Figure 7.3 *Selection of Parkesine objects (ca. 1860), including three buttons, a hair slide, a billiard ball, a paper knife, and decorative items. Parkesine—molded cellulose nitrate—was invented by Alexander Parkes of London. It was a predecessor of celluloid. (London Science Museum, Science & Picture Library)*

lonite Company in Great Britain and, later, the Zylonite Company in the United States.

By the mid-1800s, the widespread use of ivory for knife handles, piano keys, dice, chess pieces, jewelry, and billiard balls had already raised concerns that elephants were becoming endangered as a species and their ivory tusks very expensive. Significant amounts of ivory were especially consumed for billiard balls. Substitute materials, such as hard wood, were typically unsatisfactory for billiard balls on account of their poor rebounding properties. The leading American billiards supplier, Phelan & Collender, offered a prize of $10,000 in gold to any inventor who could devise a substitute for ivory.

The American inventor John Wesley Hyatt Jr., working in Albany, New York, took up the challenge and in 1869 patented the use of almost-pure collodion for coating nonivory billiard balls. But that early attempt was affected by the coating's flammability. Balls were occasionally ignited when lit cigars accidentally came into contact with them. A Colorado billiards-saloon owner, testing the collodion-coated billiard balls, wrote to Hyatt describing how hitting the balls very hard would occasionally set off a mild explosive reaction, producing a sound similar to a gunshot. He did not mind the noise, but when it happened every man in the room pulled a gun!

Hyatt—with his brother, Isaiah Smith Hyatt—continued working on the project, and in 1870 patented a process for plasticizing cellulose nitrate with camphor. They found that billiard balls made with an outer layer of the new material—which they called *celluloid*—worked exceptionally well, and their commercial venture, the Albany Billiard Ball Company, thrived. Through a separate enterprise, the Celluloid Manufacturing Company, and through licensees, many inexpensive popular items were produced, including combs, collars and cuffs, shoe horns, letter openers, and toiletware (fig. 7.4).

Hyatt's inexpensive billiard balls contributed to the popularization of the game of billiards, a game previously played mainly by the wealthy. But Hyatt never received the gold prize, and apparently never even claimed it.

Patent-infringement litigation over the use of camphor developed between the Hyatt brothers and Spill. The lengthy, complex, and expensive litigation lasted until Spill's death in 1887 and ultimate dismissal of the case by the United States Supreme Court. The case may be among the first—if not the very first—major patent disputes in the plastics industry.

It is somewhat ironic that Parkes testified on behalf of the Hyatts, rather than on behalf of Spill, his former business associate, who was at that time manufacturing Parkesine. But Parkes did not at that point have high regard for Spill, and more than anything he wanted to counter Spill's patent-right claims. After all, it was Parkes, not Spill, who had introduced the use of camphor.

Scholars still debate the merits of the legal case and who should be given primary credit for originating what came to be the first widely applied plastic. In favor of Hyatt's claim to the title of originator of celluloid is the very practical point that Hyatt's *commercial* venture succeeded whereas Parkes's did not. There is also the fact that Parkes probably viewed camphor-ethanol only as a useful *solvent* for helping liquid collodion dry to a useful material, whereas Hyatt recognized the value of camphor as a *plasticizer,*

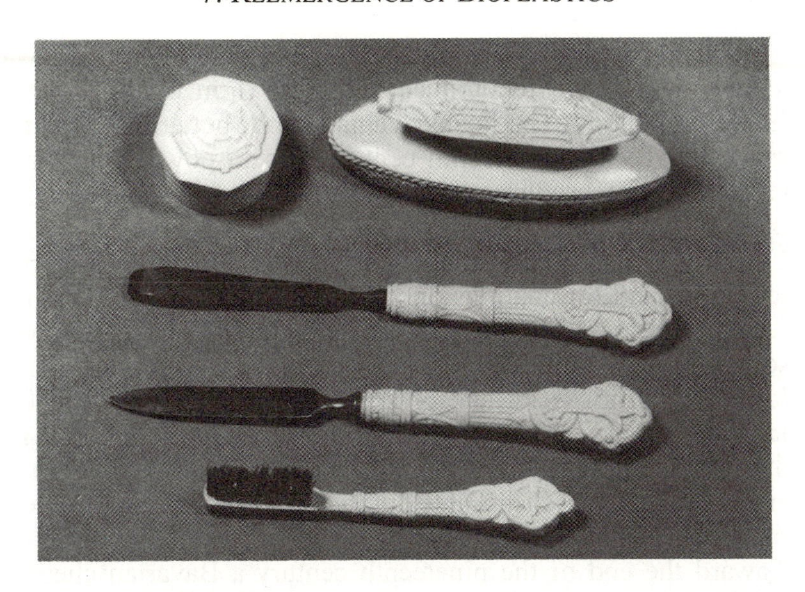

Figure 7.4 *Celluloid molded personal-grooming implements, including a powder/salve box (1"), manicure implements, and brush. The items are from the 1892 catalog of John Wesley Hyatt's Celluloid Manufacturing Company, but bear the imprint of the American Zylonite Company (1881–1890), which was acquired by Hyatt's company after extensive litigation concerning the use of camphor in nitrocellulose plastics. (Courtesy of Keith Lauer and Julie Robinson, and Collector Books)*

which made the solid moldable—a thermoplastic. On the other hand, Parkes was the first to use camphor with cellulose nitrate, and Hyatt knew of that use. Although court cases ultimately require final decisions, history does not; what is important is that the contributions of each are known.

Celluloid was used in 1882 to make photographic film, and later for production of the first motion-picture films. It was also used to make windshields for carriages and eventually for automobiles, but there was a tendency for it to yellow in prolonged sunlight. Its flammability continued to be an important limitation. Today it still has a few applications, including Ping-Pong balls and ballpoint pens.

Casein was also widely used commercially, first for paints and glues, later for plastics. Eighteenth-century farmers used buttermilk—the liquid left after churning cream for butter—mixed with lime and a colorant to produce a durable, waterproof "milk paint" for sheds and barns. Red was a popular color, partly because it was easy and inexpensive to produce with rust (iron oxide) or even with livestock blood. The Pennsylvania Dutch "buttermilk red" is a casein paint. By the nineteenth century, casein powder and paste were commercial products, available to the public and used for painting furniture and toys.

Casein was also used to make glue (it was sometimes preferred over gelatin) and for coatings on specialty papers intended to imitate cloth. Casein fiber was also manufactured. The casein was dissolved in an alkaline solution and extruded into a bath of formaldehyde, sulfuric acid, and glucose. The reaction between formaldehyde and casein formed molecular cross-links, the sulfuric acid neutralized the alkalinity, and the glucose acted as a plasticizer. The fiber was cheap and durable, and was used to extend wool, which was more expensive.

Toward the end of the nineteenth century a Bavarian chemist, Adolf Spitteler, invented moldable casein plastic, quite by accident. He kept a cat in his laboratory to keep it free of mice. One day the cat jumped up onto a lab bench and knocked over a flask of formaldehyde, which ran off the bench top and dripped into the cat's saucer of milk that was on the floor. When Spitteler discovered the accident he was surprised to find a hard waxy hornlike substance in the saucer. He immediately recognized the significance of his discovery and set about to commercialize it.

The commercial process that soon developed was based on first isolating pure casein from buttermilk or skimmed milk, which were inexpensive and plentiful waste by-products of the dairy industry. (They could not be used for hog feed or for cheese.) The casein was mixed with water and heated, colorants were added—ranging from bright colors to soft pastels—and then the mixture was extruded into a formaldehyde bath. The early extruders resembled an old-fashioned food grinder, and could produce sheets, rods, tubes, or disks. In the formaldehyde bath, cross-linking produced a solid, which hardened and cured. If alum—aluminum potassium sulfate—was used in the mixture, the resulting plastic could be machined before it had thoroughly set and hardened, making it possible to produce intricate shapes and surface designs. Casein plastics were durable and corrosion-resistant, and polishing produced a pleasant, hornlike appearance that made them popular for ornate decorative items, like buttons, beads, hat ornaments, brooches, and buckles, but they were also used for practical items—knitting needles, crochet hooks, pen barrels, knives, letter openers, and brush handles. The manufacture of casein plastics was a significant industry at the beginning of the twentieth century.

Other bioplastics were also being produced at that time. Molded horn jewelry was very popular. Shellac, extracted from harvested lac insects, was not only used for paint and varnish but, mixed with fillers, for small solid articles. Soy protein, like casein, was reacted with formaldehyde to produce plastics.

Gelatin has had a long and interesting history as a commercial material. Mixed at high concentrations with water and plasticizer, it forms a slurry or *viscose.* When dried, the viscose forms a tough, hard solid. Bottle caps and printer rollers were made from gelatin on a large scale. Gelatin printer rollers are still used for special printing applications such as etching. Gelatin is also still widely used in pharmaceuticals as an encapsulating material, and in photography.

Machinery improvements were also being made during the nineteenth century, including forerunners of modern extrusion, masticating, injection molding, compression molding, and blow molding machines.

The history of plastics changed dramatically in the early 1900s when petroleum emerged as a source of fuel and of chemicals. The early bioplastics were simply displaced by plastics made from synthetic polymers. Phenolics were commercialized in 1907 and bakelite provided nonivory billiard balls, replacing celluloid. (Nowadays billiard balls are made of other plastics; original bakelite billiard balls are valuable collector's items.) Many other plastics followed in the 1920s. World War II brought on a large increase in plastics production, a growth that continues to this day.

In the 1910s Henry Ford experimented with using agricultural materials in the manufacture of automobiles. Ford was partly motivated by a desire to find nonfood applications for agricultural surpluses, which existed then as they do now. He tried out many agricultural crops, including wheat. Coil cases for the 1915 Model T Ford were made from a wheat gluten resin reinforced with asbestos fibers. Eventually he focused on soybeans, and in the 1920s began promoting soybean products at every opportunity. He recruited Robert Boyer, a young chemist, to lead the research.

In the following few years, uses were found for *soy oil* in automobile paints and enamels, in rubber substitutes, and in the production of glycerol for shock absorbers. Viscous solutions of *soy protein* were extruded and "set" in a formaldehyde bath to form fibers for upholstery cloth. But Ford's special interest was in converting *soy meal* into plastics. Soy meal is what is left after soybeans are crushed or ground into flakes and the soy oil extracted with a hydrocarbon solvent. Soy meal is about 50 percent protein and 50 percent carbohydrate—mainly cellulose.

The compositions of Ford's soy plastics, and the methods of their processing, evolved over time and varied according to the application. In general the resin core was made of soy meal reacted with formaldehyde to produce cross-linked protein (reminiscent of casein plastics and animal horn), but for added strength and resistance to moisture, phenol or urea was cocondensed with the protein. The resulting resin was part phenol formaldehyde (or urea formaldehyde) and part cross-linked soy protein; the soy meal was not merely a filler. The condensation took place in the presence of the cellulose and other carbohydrates that were part of the soy meal. Fillers, up to 50 to 60 percent, provided additional cellulose fibers, from wood flour or pulp from spruce or pine, cotton, flax, hemp, ramie, even wheat. The final mix was about 70 percent cellulose and 10 to 20 percent soy meal. When additional strength became necessary, glass fiber was also used. Relatively low pressures and temperatures were used in the molding process.

Soy meal plastics were used for a steadily increasing number of automobile parts—glove-box doors, gear-shift knobs, horn buttons, accelerator pedals, distributor heads, interior trim, steering wheels, dashboard panels, and eventually a prototype exterior rear-deck lid. Finally Ford gave the go-ahead to produce a complete prototype "plastic car," including an entire plastic body. The body consisted of fourteen plastic panels fixed to a welded tubular frame (instead of the customary parallel I-beam frame). The panels and frame each weighed about 250 pounds. The total weight of the automobile was 2,300 pounds, roughly two-thirds the weight of a steel model of comparable size (fig. 7.5).

Figure 7.5 *Prototype "soybean plastic" automobile. Henry Ford developed plastics from soybeans and other agricultural products for use in the manufacture of automobiles. The prototype—shown here with its designer, Lowell E. Overly—was introduced in 1941 with great fanfare. The body consisted of fourteen compression-molded panels fixed to a tubular steel frame. The panels were molded with a core resin that was part phenol (or urea) formaldehyde and part cross-linked soy protein. World War II interrupted Ford's development of soybean plastics. (From the Collections of Henry Ford Museum & Greenfield Village, P189016353)*

Ford, a master at generating publicity, exhibited the prototype with great fanfare in 1941. But then, by late 1941, Ford no longer publicized the "plastic car," probably for a variety of reasons. World War II played a role: armament work took precedence over almost everything else, effort and materials had to be reallocated, and steel shortages limited all nondefense manufacturing. There were also production-related problems. Molded-plastic technology was not yet well developed, limiting design options to easy-to-mold straight-line features. And the prototype had a noticeable formaldehyde odor. Ford's 1941 soy-plastic car did not survive, and its remains have been lost. Since then the use of plastic in automobiles has become common, but the use of plastics from renewable resources got sidetracked.

One well-established bioplastic that has survived the growth of the synthetic plastics industry is cellophane, a sheet material derived from cellulose. Its manufacture begins with shredded wood pulp, which is treated with caustic soda. In water, cellulose swells but does not dissolve, because water only solvates the amorphous, noncrystalline regions. The crystalline regions of cellulose—around 30 to 40 percent in natural cellulose—are not permeable to water, and the crystalline regions keep cellulose from dissolving. Caustic soda disrupts the crystalline regions and produces a solution of alkali cellulose.

Treatment of the solution with acid generates a cellulose gel, which is washed, purified, and bleached. Softening agents—often either glycerol or ethylene glycol—and other chemicals are then added. The resulting material is cast into a sheet, which is dried and wound onto large rolls. The solid sheet material—an example of "regenerated cellulose"—is much less crystalline than the original natural cellulose, and consists largely of a tangled amorphous *felt* of cellulose molecules. The sheet at this stage therefore interacts much more strongly with water vapor or water; it has a high water-vapor permeability, and loses strength when wet, but it does have good resistance to oils and greases.

Coatings—on both sides of the sheet—are added to provide a barrier to moisture and other properties, such as barriers to gases, solvent sealability, and heat sealability; cellophane is not a thermoplastic, and the base sheet is not heat-sealable. The original coating formulation is similar to one that is still in use. It consists of four components, including a thin nitrocellulose layer, a wax moisture-proofing barrier, a plasticizer, and a blending agent. The coatings add only 10 percent to the weight of the sheet.

An alternative treatment uses poly(vinylidene chloride) to enhance barrier properties instead of the nitrocellulose-wax coating. Either coating gives cellophane a water-vapor permeability comparable to high-density polyethylene, making it suitable for packaging applications in which particularly high moisture barriers are required.

Uses of waterproof cellophane as a packaging material began in the early 1930s. With its transparent and glossy allure, and its hygienic promise of protection and freshness, it soon became a prime packaging material.

Production of cellophane peaked in the 1960s at about 750 million pounds a year in the United States, but since then it has largely been replaced by synthetic plastics such as polypropylene. Cellophane is still used in food packaging for such items as potato chips, candies, and baked goods. It is transparent and glossy, and its stiffness allows bags to stand upright. Nonfood applications include cigarette packages, cigar wrappings, and others. It tears easily when notched, making it an excellent material for easy-opening packaging applications, and it has superior printability.

Uncoated cellophane film disintegrates in ten to fourteen days and totally biodegrades in one to two months. Nitrocellulose-wax coatings also totally biodegrade, in six months or so. Coatings of poly(vinylidene chloride), on the other hand, do not biodegrade. They disintegrate to a friable powder.

Cellophane is derived from wood, which is not a rapidly renewable resource, or from cotton, which is. Technologies for regenerating cellulose from other rapidly renewable sources, such as straw, have not been commercially competitive. Cellulose can be produced from microorganisms through fermentation, but the scale of such production is considered to be limited. Bacterial cellulose, for example, is used in the diaphragms of some high-fidelity loudspeakers. Cellulose from the bacterium *Acetobacter xylinum* forms a paperlike sheet material that is much stronger than paper.

Cellulose can be chemically modified to produce a wide variety of *cellulosic* plastics—some of which are thermoplastic—and fibers. Cellulose acetate is produced by reacting cotton fiber with acetic acid and acetic anhydride, using sulfuric acid as a catalyst. It is thermoplastic, and films are obtained by extrusion or casting from acetone solution. The film is clear and strong, and has high oxygen and water-vapor permeability, making it excellent for fresh produce and baked goods—it "breathes" and does not fog. It is also resistant to both oils and greases.

Other thermoplastic cellulose derivatives are cellulose propionate, hydroxypropyl cellulose, and ethyl cellulose. Methyl cellulose and hydroxypropylmethyl cellulose are not thermoplastic, but films can be cast from solvent. Some chemically modified celluloses do not have the biodegradability characteristic of cellulose.

Before 1950, cellulose derivatives were the most important group of thermoplastics, and today cellulosic fibers still make up about 8 percent of the fiber market. Some people refer to cellulose derivatives as "semisynthetics," because they are chemical modifications of the natural material. But the polymers themselves are definitely not synthetic, and cellulosics could just as well—or even better—be referred to as "modified biopolymers."

The New Bioplastics

Growing environmental concern has revived interest in developing materials from renewable feedstocks, and new bioplastic technologies have begun to emerge. The Reading List at the end of the book provides an entry into the research that has been done in the area of bioplastics.

Some of the new technologies may not make the grade—for technical reasons. Others will prove to be technologically sound, but may not be able to attract the substantial amounts of investment capital needed to develop commercial products. But some are well on their way to commercial success. We are at the exciting stage of witnessing the birth of these new technologies, and we can expect a growing number of products to reach the marketplace.

There are three important types of new bioplastic material. They differ according to the means of commercially producing the resins from which the bioplastics are processed. (Chapter 6 describes the structures, origins, and nonplastic commercial uses of the various raw materials referred to here.)

(1) Plastics processed from polymers extracted directly from their natural origin

There are many ready-made polymers in nature available for use in manufacturing bioplastics. Starch is the prime example. It is important not only because it is the least expensive biopolymer but because it becomes thermoplastic when properly plasticized with water or other plasticizers. Starch formulations can therefore be processed by all of the methods used for synthetic resins. For example, they can be film extruded, injection molded, and thermoformed.

The limitations of starch formulations have been in their physical properties: poor water resistance and modest strength. Various strategies have been employed to get around these problems. In one of the earliest uses of starch in plastics, it was combined with polyethylene—typically to the extent of around 6 percent starch—to form films. The strategy was to have biodegradation of the starch component lead to rapid fragmentation of the plastic film, and possibly hasten degradation of the polyethylene. Photocatalytic additives aimed at further enhancement of degradation in sun-exposed applications.

Starch-polyethylene formulations with higher starch contents, of 20 to 80 percent, have been developed as well. The acceptance of such formulations depends on the intended application, and on whether the polyethylene residue is tolerable and nontoxic. As agricultural covers they have the advantage that they need not be collected at the end of the growing season, saving labor and disposal costs. That advantage has to be weighed against the effect of possible accumulations of nondegraded residue in the soil. Similarly, as compost bags with 23 to 25 percent starch, they have the advantages of low cost and "curb sturdiness," advantages that have to be weighed against the possible presence of polyethylene residues in the compost for some indeterminate time. Questions of the final environmental fate and ecological effect of the polyethylene residues continue to be the subject of a considerable amount of study (see Chapter 5).

Starch can be blended with poly(vinyl alcohol). Like polyethylene, poly(vinyl alcohol) is a synthetic polymer but, unlike polyethylene, it is totally biodegradable. The blends are thermoplastic and can be processed by extrusion, injection molding, blow molding, film blowing, and thermoforming. Products can be made that do not dissolve in cold or hot water, but swell to degrees that depend on additive composition and processing. There is some sensitivity to humidity. Their mechanical properties at 55 percent relative humidity resemble polyethylene (fig. 7.6).

The blends biodegrade at rates that depend on composition and crystallinity. In degradation tests in activated sludge, more than 90 percent of some samples degrade in ten months, according to measurements of weight loss and carbon dioxide production.

Figure 7.6 *Biodegradable blown films made from gelatinized starch and poly(vinyl alcohol). (Courtesy of R. Shogren, U.S.D.A.)*

Water-soluble blends can be made, with the solubility depending on the amount and molecular weight of the poly(vinyl alcohol) component and on its crystallinity. Crystallinity varies according to the level of residual acetate groups that remain following the production of poly(vinyl alcohol) from the alcoholysis of poly(vinyl acetate) (Chapter 4). One application for such blends is making laundry bags for hospitals and other institutions, where the bags dissolve during the washing and biodegrade after disposal into sewage.

Starch-polycaprolactone blends have also become important commercially. Polycaprolactone, like poly(vinyl alcohol), is a synthetic polymer, but it too is totally biodegradable. The resulting blends have significant strength; their specific properties depend on the relative amounts of starch and polycaprolactone. Thermoplastic starch sheeting can also be laminated with a polycaprolactone coating to make it water resistant (fig. 7.7).

Starch-based materials have been developed for a number of commercial applications, and several major companies produce starch-based resins. The resins have been used to manufacture agri-

Figure 7.7 *Preparing thermoplastic starch sheet laminated with a water-resistant coating of polycaprolactone by coextrusion. (Courtesy of R. Shogren, U.S.D.A.)*

cultural covers; compost bags and trash bin liners; household items, such as disposable bowls, eating utensils, and straws; single-use disposable packaging film; diaper backings; disposable golf tees; and personal hygiene articles, like combs and disposable razors.

Extruded starch foam is used as a loose-fill packaging material in the form of pellets or "peanuts." They are water soluble and biodegradable, yet have the resilience and compressibility of polystyrene. The material is processed under carefully controlled conditions of water content and extruder die temperature. Flash generation of steam as the starch leaves the die produces an expansion of the material to form a *cellular* structure, made up of numerous cells distributed throughout the material. Starch foam packaging peanuts now make up about 20 percent of the foam packaging market. Blending starch with poly(vinyl alcohol) increases the water resistance of the foam. The foam can then be made into shaped articles such as cups and plates (fig. 7.8).

Figure 7.8 *Starch-poly(vinyl alcohol) foam articles for a variety of packaging and serving needs. (Courtesy of R. Shogren, U.S.D.A.)*

Starch-based materials that include significant amounts of petroleum-derived polymers are not bioplastics as the term is used here. The technologies for producing them have been important not only because they were among the first aimed at making programmed-degradable plastics, but because the abundance and low cost of starch make the production of starch-based bioplastics stand as a continuing technological goal.

Starch foam can also be made more water resistant without using a synthetic polymer, by derivatization—a chemical modification in which one reactive group in a molecule is replaced with another, giving more desirable properties. For example, starch can be acetylated, whereby some of the hydroxyl (–OH) groups are converted into acetyl (–$OCOCH_3$) groups. The acetylated starch foam has a higher water resistance but remains biodegradable (fig. 7.9). Alternatively, starch foam can be coated with a layer of acetylated starch, but the best results are obtained when acetylated starch is used to coat a starch-acetylated starch blend.

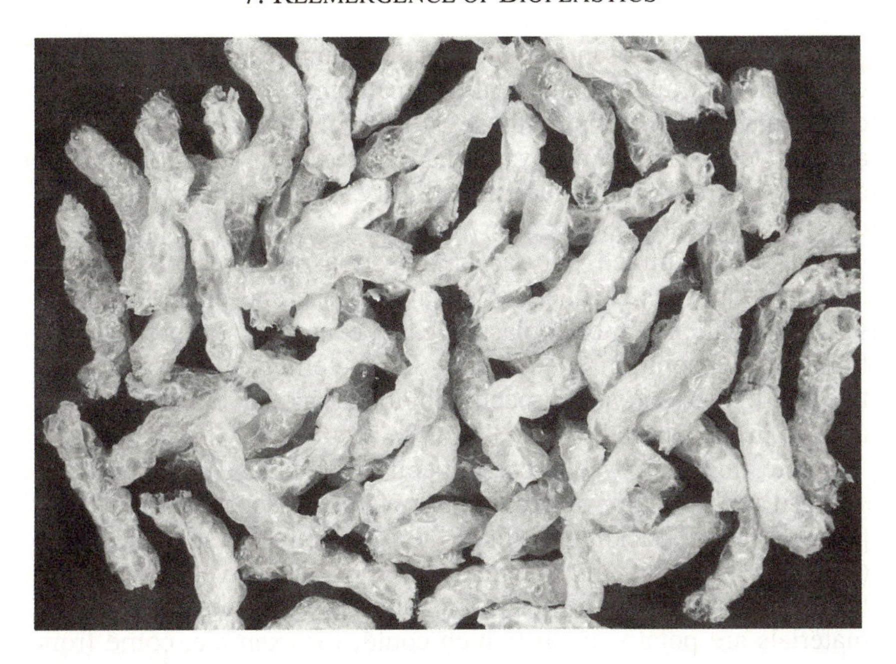

Figure 7.9 *Water-resistant foam packaging "peanuts" made by extruding starch acetate and water. (Courtesy of R. Shogren, U.S.D.A.)*

Today, there is a particularly exciting—but formidable—challenge of discovering a commercially viable biodegradable starch-based bioplastic made entirely from renewable raw materials and having adequate properties for a wide variety of applications.

In addition to starch, other abundant polysaccharides extracted directly from their natural sources include chitin, cellulose, agar, carrageenan, pectin, and alginate.

Chitin and chitosan materials have reached commercial production levels. Importantly, they are not thermoplastic but can be prepared as films by evaporation of solvent for a variety of film and fibers product applications. Films and fibers have good mechanical properties and films have low oxygen permeability. Chitin and chitosan are also used as cellulose additives to give increased wet-strength to paper towels, shopping bags, and diapers. Blends of chitosan and poly(vinyl alcohol), plasticized with sorbitol and sucrose, have been studied in the laboratory.

Chitin and chitosan have biomedical applications as well, in contact lenses and wound-healing treatments. Fibers of chitin or chitosan, blended with cellulose, silk fibroin, collagen, or glycosaminoglycans (including hyaluronan), have been studied with the aim of improving the biocompatibility of the fibers in wound dressing applications.

The use of mixed polysaccharides is a more recent development. Composites of finely divided cellulose and chitin or chitosan, although not thermoplastic, can be molded or prepared as films by evaporation of solvent. Starch-cellulose, starch-derivatized cellulose, and starch-pectin combinations have also been examined in the laboratory.

Interesting starch-cellulose-mineral composites have been developed with the aim of providing low-cost, disposable, and compostable food packaging, like the distinctively shaped boxes ("clamshells") that are so widely used in the drive-through and take-out services of the fast-food industry. The inexpensive raw materials are potato starch (which could, for example, come from the potato peel residues of french-fries production), cellulose (from wood pulp), and calcium carbonate (from limestone). The clamshells have an outer coating of poly(vinyl alcohol) and glycerol, and an inner waterproofing coating of modified paraffin wax. These novel biodegradable composite materials might be said to bridge plastics and ceramics.

Abundant proteins include soy protein, zein, gelatin, and casein. With the renewed interest in biodegradable plastics, awareness of the value of soybeans has been revived. Soy protein, with and without cellulose extenders, can be processed with modern extrusion and injection molding methods. Mixtures of soy protein and starch are also thermoplastic and can be extruded and injection molded. Even soy protein-clay composites have been examined for possible applications in packaging for the fast-food industry.

Zein is thermoplastic and water insoluble. It resists microbial attack and forms fibers that are strong, washable, dyeable, and grease resistant. Cloth products manufactured from zein fibers have already reached the marketplace. Gelatin is now widely used for drug and vitamin encapsulation, and for other miscellaneous biomedical applications, including artificial skin. It is, however, thermoplastic and is a potential feedstock for a wide range of plastics.

Many water-soluble biopolymers such as starch, alginate, pullulan, gelatin, soy protein, casein, zein, wheat gluten, and whey protein are known to form flexible films when properly plasticized. Thin films are often applied to foods where they act to preserve flavor, maintain optimum moisture content, enhance appearance, extend freshness, and provide protection. Even home bakers are familiar with the use of cornstarch to produce a glaze on breads and rolls during the last few minutes of baking.

Although such films are regarded mainly as food coatings, it is recognized that they have potential use as nonsupported standalone sheeting for food packaging and other purposes. Their mechanical properties depend on plasticizer content and relative humidity.

Polysaccharide-protein composites that have received attention include starch-casein, starch-gelatin, starch-zein, and starch-soy protein combinations. They are all readily biodegradable. Starch-protein compositions are especially interesting because they satisfy the nutritional requirements for farm animals. Hog feed, for example, is recommended to contain 13 to 24 percent protein, complemented with starch. Used food containers and serviceware collected from fast-food restaurants could be pasteurized and turned into animal feed—a rather extraordinary example of recycling waste.

Polysaccharides and proteins are already in large-scale commercial production. They are available now for use as bioplastics feedstocks. Starch stands out in terms of its low cost (it is competitive with polyethylene) and its thermoplasticity. Its limited physical properties can be improved through chemical modifications, through coatings, and through use in combination with other biopolymers in blends and composites. Other polysaccharides and proteins are, nevertheless, also potential feedstocks for new, useful plastics.

(2) Plastics processed from polymers produced commercially in large-scale fermentation processes

Through fermentation processes it is possible to induce microorganisms to produce biopolymers on large scales. The substrates fed to microorganisms in the commercial fermentation processes are generally naturally occurring and of plant origin.

Polyhydroxyalkanoates—naturally occurring polyesters—have already played an important role in the commercialization of bioplastics. Polyhydroxybutyrate (PHB) plastics are brittle, but copolymers of hydroxybutyrate and hydroxyvalerate (PHBV) display a range of brittleness, strength, and other properties according to the copolymer composition.

The ratio of feedstocks—glucose and propionic acid—determines the composition of the copolymer. The polymer material, in the form of granules, is removed from the cells by aqueous extraction, resulting in a white powder. Depending on the hydroxyvalerate content, PHBV can display physical properties and processing behavior resembling polyethylene or polypropylene, and can range from a brittle plastic to an elastomer. PHBV properties can be further modified with additives.

PHBV is thermoplastic and can be processed by injection molding, extrusion blow molding, film and fiber forming, and a variety of coating and lamination techniques. It has been processed into a wide range of packaging materials, including bottles, toiletry articles such as disposable razors and combs, eating utensils, dishes, cups, credit cards, plant pots, golf tees, motor oil containers, and other products. An early commercial application was in the manufacture of a blow-molded bottle for the shampoo industry. Hair shampoos are now usually biodegradable, and a biodegradable bottle is a fitting package.

Significantly, PHBV has substantial water resistance, greater than most polysaccharides and proteins. Stored in humid air PHBV is stable, and its oxygen barrier and aroma barrier properties are not sensitive to relative humidity. In coated paper products PHBV can be used in place of polyethylene. When applied to starch foamed articles PHBV coatings can provide resistance to hot and cold water; an adhesive layer of shellac, a natural resin, prevents delamination (fig. 7.10).

PHBV is biodegradable in soil, river water, seawater, aerobic and anaerobic sewer sludge, and compost. For example, PHBV mineralizes in anaerobic sewer sludge to carbon dioxide and methane to the extent of nearly 80 percent in thirty days. Compression-

Figure 7.10 *Starch foam cup and tray coated with water-resistant PHBV. An adhesive layer of shellac, a natural resin, prevents delamination of PHBV from the starch. (Courtesy of R. Shogren, U.S.D.A.)*

molded into dog-bone shapes, it loses 60 percent of its weight after six months in a municipal leaf compost. Its biodegradation rate, like its physical properties, depends on copolymer composition, molecular weight, degree of crystallinity, surface area, and the presence of biodegradable additives such as plasticizers.

Biodegradation begins with bacteria or fungi colonizing the surface and excreting an extracellular depolymerase enzyme that degrades and solubilizes the polymer near the cell. Fragments are then absorbed through the cell wall and mineralized. At elevated temperatures, hydrolysis also contributes to degradation. Moreover, the formulation of PHBV makes it compatible with recycling and clean incineration.

PHBV has also been widely tested for biomedical uses. It is not only biocompatible but thermoplastic, so it can be molded into any desired shape. It biodegrades in the body solely by hydrolysis, which results in slow degradation rates—on the order of several months to a year or more. The degradation product—hydroxybutyric acid—is a mammalian metabolite and occurs in low concentrations in humans. The applications that have been considered or are under development include drug delivery systems, sutures, staples, screws, clips, fixation rods, implants, and others.

The future of large-scale production and marketing of PHBV products is not clear, but the excellent properties of PHBV make it a strong candidate for eventual commercial success.

Pullulan, a polysaccharide, is also prepared commercially, through yeast fermentation. It can be processed into a bioplastic that is hard, strong, tough, and elastic. Pullulan has good oxygen barrier properties and has been developed for food-packaging applications. Fibers have been drawn from concentrated solutions.

Bioplastics prepared with fermentation technologies typically have excellent physical properties. Their limitation is currently their cost. They are promising materials nevertheless, because the fermentation technologies that are now being developed may allow very large scale production levels and the accompanying lowered costs.

(3) Plastics processed from resins polymerized from naturally occurring monomers: "honorary biopolymers"

In a third category of bioplastics are materials polymerized from naturally occurring low-molecular-weight biomolecules. Polymerizations are carried out in some cases to produce thermoplastics, in others to produce sturdy thermosets.

Poly(lactic acid)

One important example is poly(lactic acid) (PLA), a polyester. The substantial commercial production of poly(lactic acid) has been made possible by recent developments in large-scale fermentation technologies by which the monomer, lactic acid, can be produced (see Chapter 6).

Poly(lactic acid) is thermoplastic and can be processed by most common methods; it can be sheet and film extruded, injection molded, blow molded, thermoformed, and fiber spun. It can be made flexible or rigid, and it is inherently clear but can be processed to be opaque. It can accept fillers, and in some applications its high strength allows for down-gauging—the use of thinner than usual sheets.

Importantly, PLA is insoluble in water and has good moisture and grease resistance. Its mechanical properties can be modified by varying its molecular weight and its crystallinity. Its properties can also be modified by copolymerization of the lactic acid with glycolic acid or caprolactone.

Poly(lactic acid) is biodegradable in seawater and other environments, and is compostable. It degrades by hydrolysis, which can proceed even in the absence of enzymes. It can therefore be recycled back to the monomer, lactic acid.

Poly(lactic acid) resins are being developed and marketed through a number of commercial efforts. They are used for agricultural mulch film; compost bags for yard trimmings, food waste, and other compostables; and bin liners for food scrap collection. Its clarity, high gloss, and stiffness also make it useful for recyclable and biodegradable packaging, such as bottles, yogurt cups, and candy wrappers. For the fast-food and institutional-food-service markets, it has been used to make disposable single-use items like plates, cups, cutlery, covers, straws, and film wrap.

Other nonpackaging products include sporting and recreational products such as golf tees; coatings for paper and cardboard; and fibers—for clothing, carpets, sheets and towels, and wall covering.

Poly(lactic acid) has been approved for, and is used in, a number of biomedical applications. Poly(lactic acid-*co*-glycolic acid) has been used, for example, to fabricate microcapsules, microspheres, and nanoparticles for drug delivery systems. In one process, the polymer is first dissolved in an organic solvent to form a homogeneous solution. A bioactive agent—like hydrocortisone, progesterone, or taxol—is dissolved or dispersed in the solution and the organic solvent is removed to provide a solid particle, insoluble in water.

A large amount of research is currently under way to develop specific delivery systems. For example, delivering heparin to blood vessel walls by oral ingestion is hindered by its easy degradation into inactive fragments, and administering it into the blood stream requires continuous infusion. A novel local delivery system uses poly(lactic acid-*co*-glycolic acid) microspheres to encapsulate the heparin; the microspheres are then sequestered in an alginate gel. Controlled release of the heparin can be maintained over a period of twenty-five days.

Another system is being developed for delivering anticancer drugs into the brain by injection, for treatment of cerebral tumors. Microspheres of poly(lactic acid) encapsulating the drug 5-fluorouracil are coated with chitosan. After an initial rapid release of 25 percent of the drug, the release continues at a slower rate for a period of thirty days.

Bone repair materials have been targeted as another valuable application. In the United States over a million operations a year are performed involving bone repair, but current reconstructive methods are not without limitations. A bone graft from the patient (autograft) causes some donor-site morbidity and is limited in supply. A bone graft from another source (allograft), usually cadavers, risks rejection and disease. Metal plates or rods sometimes result in stress shielding—the surrounding bone tissue is shielded from normal load-bearing stress and is reabsorbed by the body, leading to loss in bone density and the possible need for further surgery.

Bioresorbable plastics provide an alternative orthopedic material for use in bone regeneration applications. One composite being studied contains the copolymer, poly(lactic acid-*co*-glycolic acid). The material supports the growth and adhesion of new cells, and can be made porous to provide a large, continuous surface for cell proliferation throughout the matrix. The degradable material maintains mechanical integrity while the bone heals itself. A poly(lactic acid) implant will degrade completely in a time that depends on its size and shape, but generally in less than twenty-four months.

In another use, poly(lactic acid) containing growth factors has been applied as a coating to metal implants used to stabilize bone fractures. Eighty percent of the growth factor is released within forty-two days, and improved healing is observed compared to an uncoated implant. Interestingly, even a poly(lactic acid) coating with no incorporated growth factor showed some improvement.

Poly(lactic acid) has many excellent properties and is suitable for a wide range of applications. Its future looks promising.

Polymers of Triglycerides

Plant and animal triglycerides have become the basis for a new family of sturdy, durable composites that have long useful lifetimes. The triglycerides are first converted to a more chemically reactive form capable of polymerizing—epoxidation is one method. The initial liquid resin is a low-molecular-weight polymer that is then combined with catalysts and accelerators to facilitate the cross-linking reaction. It is injected into a mold containing a reinforcing fiber and then heat-cured in the mold to form a rigid thermoset. Genetically engineered oils, with increased and otherwise unattainable levels of fatty acid unsaturation, improve the properties of the final composite.

Soy oil resins with glass fiber reinforcement can be made into long-lasting, durable thermoset materials. Practical applications of these sturdy materials include agricultural equipment, the automotive industry, construction, and other areas (fig. 7.11).

These new vegetable oil composites are only vaguely reminiscent of Henry Ford's soybean plastics; they are significantly improved versions. Ford's plastic involved cross-linking the soy protein that remained in the soy meal after the oils had been extracted. For strength, Ford's plastic had to have a substantial amount of phenol-formaldehyde or urea-formaldehyde resin. The new soy oil composites are based only on renewable raw materials, and are nevertheless strong. They are also free of formaldehyde.

Plant oils other than soy oil, and fibers other than glass, can also be used in the process. Plant fibers from jute, hemp, flax, wood, and even straw or hay are alternatives—soy-based resins have a strong affinity for natural fibers. If straw could replace wood in compression-molded composites now widely used in the construction industry—such as fiberboard— it would provide a new use for an abundant, rapidly renewable agricultural commodity and at the same time conserve less rapidly renewable wood fiber.

In the future it might be possible to use cellulose nanoparticles in place of glass fibers to produce strong, durable—and biodegradable—composites.

Figure 7.11 *Prototype side panel for a hay baler (8' x 3', 25 lb.). A foam core is enclosed on both sides with a thermoset composite made from soybean oil resin reinforced with glass fiber. The panel was recognized with a 1998 Innovations in Real Materials Award sponsored by the International Union of Materials Research Societies. Plant oil resins can also be reinforced with natural fibers from hemp, flax, straw, or other plants to produce optionally biodegradable composites. (Courtesy of University of Delaware Center for Composite Materials)*

Epoxidized soybean oil has also been polymerized with citric acid to form a coating on kraft paper, producing a biodegradable agricultural mulch. The coating increases the wet strength of the paper and reduces its degradation rate so that the mulch can inhibit weed growth for more than ten weeks (fig. 7.12). Without the coating the paper degrades after about six weeks.

Technologies for all three types of bioplastic are under continuing development, and commercial products have begun to appear

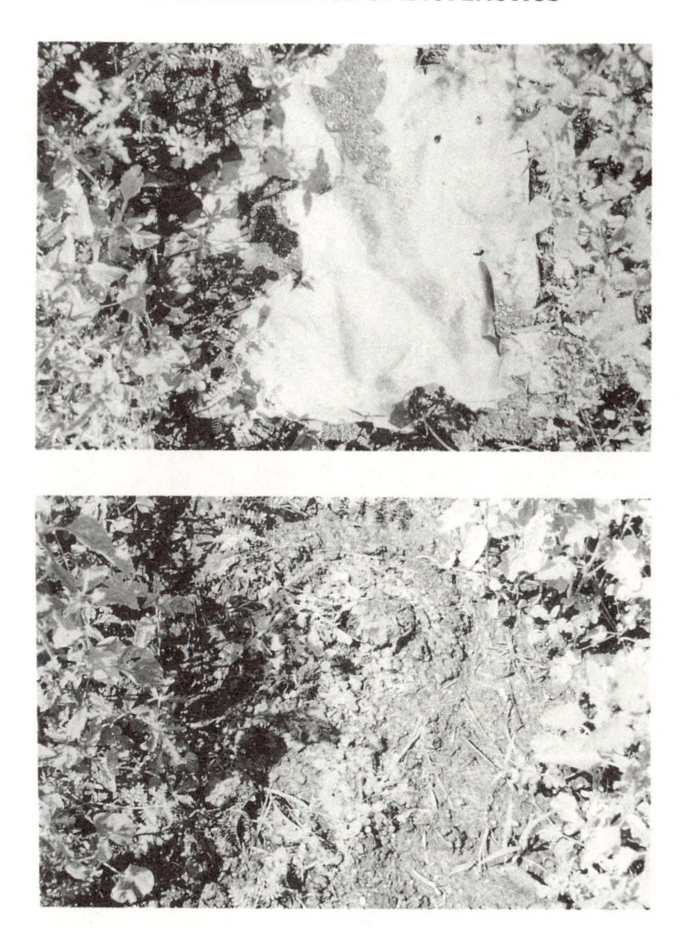

Figure 7.12 *Biodegradable mulch—kraft paper coated with a polyester made from a reaction product of epoxidized soybean oil and citric acid. The coating reduces the degradation rate and increases wet strength so the mulch can inhibit weed growth for more than ten weeks. (Courtesy of R. Shogren, U.S.D.A.)*

on the market. Several types of compost bags, for example, are already commercially available. Starch, microbial polyesters, poly-(lactic acid), and other thermoplastic resins are playing especially important roles in the emerging bioplastics industry because they can be processed with the conventional equipment of the plastics industry.

Starch-polyester blends, including starch blended with the PHBV copolymer and starch blended with poly(lactic acid), are

Figure 7.13 *Plastic eating utensils. The utensils on the left are made of 55 percent cornstarch and 45 percent poly(lactic acid). They are biodegradable and compostable. For comparison, nondegradable polystyrene utensils are shown on the right. (Courtesy of R. Shogren, U.S.D.A.)*

being examined, with the aim of maintaining the excellent physical properties of the polyesters while reducing cost—and producing a true starch-based bioplastic (fig. 7.13).

Cast-film technologies are also continually being developed, which will allow further commercialization of nonthermoplastic biopolymer resins made from naturally abundant biopolymers such as cellulose, chitin, chitosan, and others.

And the development of reinforced vegetable-oil thermoset composites creates the possibility of low-cost durable materials with biodegradability as an option.

Many formulations have been examined, and there is now an emerging bioplastics industry. The possibilities are vast, and there may be many more formulations that have commercial usefulness. But what is the potential for significant growth in the use of bioplastics? The best starting point for considering that question is an examination of some of the factors favoring and impeding the growth of their development and commercialization.

8

Factors Affecting Growth

What does the nascent bioplastics industry have on its side—what favors its good fortune? Clearly, it is the environmental friendliness of bioplastics that gives them their value and allure, a benignity that flows from three sources. Their raw materials are abundant and renewable. Their technologies are "low impact." And they are biodegradable.

Conversely, what does the industry have to cope with? Equally clearly, it is the need to succeed in the marketplace. Bioplastics have to meet performance requirements based on their physical properties, and they have to be cost competitive. All these factors are interrelated and the industry will be looking to "accentuate the positive and eliminate the negative."

Biomass Raw Materials

The growth of a bioplastics industry is favored by the ready abundance of natural feedstocks. Many agricultural and marine biopolymers are already being extracted from their natural sources on large scales, and are available commercially through mature industries. These industries have grown because practical uses have already been found for the natural polymers.

Cost is one measure of commercial availability; the more available the raw material is, the lower its cost. Commercial availability reflects how many useful applications have been found for the raw material, and how easily it is prepared from biomass.

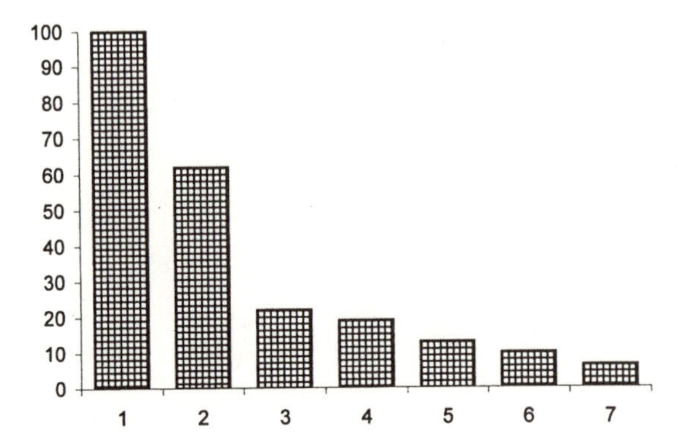

Figure 8.1 *The "inverse cost" of materials that are extracted directly from their naturally occurring sources, relative to starch: (1) starch, (2) cellulose, (3) casein, (4) chitin, (5) soybean oil, (6) gelatin, (7) agar*

Figure 8.1 shows this measure of availability for some raw materials that are extracted directly from their naturally occurring sources, as indicated by the *inverse of cost*. Starch is by far the least expensive; the world market for starch is over 70 billion pounds a year. Cellulose production is more than twice that of starch, well over 150 billion pounds a year, but it is more expensive because of the costs of extracting it from wood.

Commercial availability does not directly reflect natural abundance. Most biomass simply returns to the ecosystem through natural processes of decay, without being exploited for practical use. Cellulose is a good example. It has been estimated that 75 billion tons of cellulose are biosynthesized each year, with most simply disappearing through natural decay. Chitin is another example. A billion or more tons of chitin are synthesized in the biosphere each year, but the world market for chitin is only 1,000 to 2,000 tons a year; the rest naturally degrades for lack of alternative uses. The amount of commercially produced seaweed polysaccharides, like agar and carrageenan, is only a fraction of the total amount available. Similarly, only a small part of the animal hides generated annually is used for the extraction of gelatin.

Straw, an abundant agricultural by-product, would be a cheap source of cellulose if a sufficiently efficient extraction technology could be developed. Millions of tons a year simply degrade naturally. Corn stover—the stalks, leaves, and husks of corn—is another large source.

Fermentable sugars are also part of the renewable biomass. They are the feedstocks for the microbial production of biopolymers—like the microbial polyesters—and of polymerizable monomers—including lactic acid. Sugars can be produced either directly from plants, such as sugar beets and sugarcane, or from the hydrolysis of polysaccharides, like starch. Major starch feedstocks include corn, wheat, sorghum, and potatoes. Agricultural wastes are another, largely untapped source; they include crop residue, potato processing residue, sugar beet and sugarcane molasses, apple pomace, and others.

As additional practical uses are found for these materials, the increased demand for them will provide an incentive for greater commercial exploitation. There is room for growth because commercial availability is not now close to being limited by natural abundance.

These natural resources are, moreover, constantly being replenished, and with the exception of cellulose derived from wood, they are being replenished rapidly. Even in the case of cellulose the rate of replenishment could be increased simply through increased forestation. It is this renewability of feedstocks that surmounts one of the biggest drawbacks of current plastics; current plastics are made from resources that are being drained faster than nature can put them back. The use of renewable resources for bioplastics is a **sustainable** activity, whereas the current methods of producing conventional plastics are not.

Benign Technology

The growth of a bioplastics industry is also favored by its benign technology—a technology that has fewer environmental impacts than the technologies of a great many other industries. Much of it is water-based so that hazardous organic solvents are often not needed, eliminating not only the atmospheric pollution associated with using them, but also the fire and toxicological risks. Processing often involves few or no toxic chemical reagents. There are no harmful by-products, and there are usually no waste organic solvents to dispose of.

The production and processing of bioplastics are carried out at relatively low temperatures and pressures. Energy costs are sometimes lower, and safety hazards are reduced. For example, the production of microbial polyhydroxyalkanoates involves a one-pot fermentation process, in water, at approximately room temperature. In the case of poly(lactic acid) the monomer is produced by fermentation, and then is converted to the polymer by a solventless process.

Additives used with synthetic polymers have occasionally come under fire on account of their toxicity. Phthalates, used to make poly(vinyl chloride) soft and pliable, are the latest additives to become controversial. The level of leaching is low and in several studies has been judged to be too low to cause harm, but there is disagreement in the interpretation of the data. Concern persists especially when poly(vinyl chloride) is used to make toys for children under three years of age, including teething rings and rattles designed to be chewed. Another concern is the poly(vinyl chloride) used in medical intravenous bags. Manufacturers are looking for bioplastics made from starch and edible oils that would be suitable replacements.

These ecological advantages come with economic advantages. Water is the cheapest industrial solvent. Both raw materials and manufactured products are recyclable, so that processing waste can be efficiently reused.

Biodegradable Products

Bioplastics are nontoxic, recyclable, and biodegradable, and thereby have minimal impact on waste management. They can be formulated to be safely incinerated, and through composting they can be returned to the ecosystem harmlessly in a carbon dioxide neutral process. Composting is an especially valuable avenue of waste management in Europe and Japan where land is scarce and high landfill fees are common.

In some applications biodegradability may be the whole point of the product. The biodegradability of agricultural covers is one example. The covers need not be collected and disposed of as do conventional plastic covers. Their biodegradability plays an intrinsic functional role, and the period of use coincides with the period of degradation. Biodegradable agricultural covers also allow the controlled release of active ingredients aimed at the control of soil pests.

Another example of growing importance is the use of collection bags for yard waste, food waste, and other compostable materials. Compostable bags eliminate the need for bag removal, eliminate the deterioration of the compost from imperfect bag separation, and eliminate the cost of used-bag disposal.

Yet another example is the use of bioplastics as root system wrappings. When conventional wrappings, often of burlap or non-degradable plastic, are used to protect plants during shipping and during storage on the retail lot, the wrap has to be sturdy enough to keep the root ball intact; but if the root system is delicate, as it often is, the wrap cannot be so impenetrable as to damage the growing roots. During replanting, the wrap either has to be removed or slashed to allow the growing root system to extend beyond the ball size. A biodegradable wrap material, however, has the advantage of degrading over a controlled period of time. It might have fertilizer additives for the controlled release of nutrients. The biodegradability becomes an intrinsic part of the product application; it becomes a functional requirement as well as an environmental asset. For all such applications, the biodegradability of bioplastics will give them a product performance capability that cannot be matched by alternative products.

Another significance of biodegradability and compostability is beginning to be more appreciated, through the growing realization worldwide that further increases in agricultural productivity—required for the rapid population growth we are experiencing—cannot be taken for granted and may become increasingly difficult. The perceived importance of biodegradable, compostable materials has been shifting from their waste management value to their role in sustaining agriculture—by increasing crop yield, improving crop quality, decreasing dependence on chemical fertilizers and pesticides, and reducing water requirements. Compostable plastics not only diminish what goes into landfills; they retain value when they are returned to the farmlands where they began—closing the loop.

Properties

Bioplastics have to possess adequate physical properties. But the same physical properties of biopolymers that make them environmentally attractive feedstocks can affect their performance as materials. These properties have to be managed and controlled with technological means through the development of adequate formulations and processing.

One property of many biopolymers—including some polysaccharides and proteins—that is partly responsible for their degradability is their relatively strong interaction with water. Biopolymers are often *hydrophilic,* as opposed to *hydrophobic,* and some are even soluble in hot water.

The hydrophilicity of starch is especially significant. Starch is very inexpensive, and also thermoplastic—two important assets. Its hydrophilicity is the major barrier to its becoming a dominant bioplastic material.

Cellulose and chitin, for instance, are insoluble. A barrier in those cases is that they are not thermoplastic, so they cannot be processed by most plastics processing methods; they can, however, be cast as films. A few proteins, like casein and zein, are insoluble *and* thermoplastic, but they are significantly more expensive than starch. Bacterial polyesters and poly(lactic acid) are thermoplastic and have good physical properties; in those cases, too, cost is a more important barrier than properties.

The water resistance of starch and other hydrophilic biopolymers can be improved through technological means—nature provides many instances of composites that are strong, resilient, and water resistant. For example, chemical modifications of biopolymers can be sought which increase water resistance but do not significantly decrease biodegradability. Acetylated starch is but one example (see fig. 7.9). Starch has also been chemically modified by reaction with caprolactone; the reaction product is thermoplastic without the need for plasticizers that lead to unwanted water uptake. Proteins have a large number of different reactive side chains, which makes them well suited for a wide range of chemical modification. For instance, hydrophobic groups can be introduced by reaction with alcohols or aldehydes. Starch-zein mixtures have been made more water resistant through chemical modifications.

A combination of chemical modification *and* cross-linking can significantly improve the material properties of protein plastics. But formaldehyde is no longer acceptable as a cross-linking agent, and current research is aimed at discovering effective cross-linking agents that are more environmentally benign.

Additives or coatings are commonly used with synthetic plastics, and additives or coatings that increase water resistance can be developed for bioplastics when needed (see fig. 7.7 and fig. 7.10). Naturally occurring substances with potential use as biodegradable, moisture-proof coatings are oils, waxes, and even shellac, known since ancient times. The base cellulose sheet in cellophane, although insoluble in water, absorbs and loses moisture readily; it is not a good moisture barrier. Cellophane is made moisture proof with a coating of a nitrocellulose-wax blend. Starch films coated with a nitrocellulose-wax blend have been shown to be similarly moisture proof.

Biopolymer composites or laminates, or both, can also provide materials with increased water resistance. The use of biopolymer-biofiber composites, in particular, can alleviate one disadvantage of poor water resistance, which is low wet strength. Cellulose has been used both to be an extender and to increase water resistance. Combining the use of composites and laminates with the use of environmentally benign coatings may lead to materials with whatever degree of water resistance is required, while retaining biodegradability.

In specialized applications, the solubility of some bioplastics in hot water can even be exploited to make their solubility part of the product use. Starch-poly(vinyl alcohol) plastics, for example, are already being used to make water-soluble laundry bags for hospitals. The bags containing soiled or contaminated clothing are placed directly in the washing machine, where they dissolve. Or imagine a frozen-food product packaged in a pouch that is soluble in hot water. When dropped into a pot of boiling water, the pouch would simply become part of the recipe! No waste. No need for the clumsy emptying of pouches.

Tensile strength is a measure of the force required to break a test specimen when it is placed between two clamps and drawn. (See more on tensile strength in the Appendix.) A typical commercial plastic wrap for general household use might have a tensile strength of several thousand pounds per square inch (psi). Some bioplastic formulations have limited tensile strength, especially when plasticized sufficiently to yield a flexible and pliable product.

Many applications may not require more than moderate strength. Sandwich bags need not be as strong as trash bags. If a tensile strength of 2,000 psi will do the job for a given application, a large excess may not be necessary. But if an increased tensile strength is needed, there are several ways to achieve it.

As with water resistance, environmentally benign chemical modifications of biopolymers can be developed that increase tensile strength. The tensile strength of wheat gluten can be increased fivefold by first introducing additional reactive groups, then cross-linking the protein chains. Other routes to improving strength are introducing additives, and using composites of biopolymers and natural fibers, as in the fiber-reinforced composites found in nature. Biopolymer-biofiber composites could provide a route to strong materials without reducing their ability to biodegrade. With a combination of cross-linking *and* fiber additives, protein plastics have been made that possess mechanical properties similar to polystyrene or poly(vinyl chloride). In some formulations under development, cellulose is used to increase both tensile strength and water resistance.

Natural biopolymer fibers, like cotton, derive their strength from the arrangement of the biopolymer chains within the fiber. As more becomes known about the arrangement of the biopolymer chains in bioplastics, it may become possible to discover means of exerting greater control over that arrangement. Means for control might be achieved through the development of formulations, or the optimization of the methods of processing, or both. For example, the degree of crystallinity often affects physical properties, and means of controlling the degree of crystallinity are being sought.

A growing number of scientists are now working to develop practical methods for increasing the strength of bioplastics as well as their resistance to water, without introducing deleterious effects on their biodegradability.

A third concern over performance is the susceptibility of bioplastics to unwanted premature attack by microorganisms—such as those in fungal growth—and by macroorganisms—such as insects or vermin. The very organisms that are counted on to degrade bioplastics after use cannot be counted on to ignore the bioplastics until they are discarded. Biopolymers vary with respect to how susceptible they are to microorganism damage. Environmental conditions such as temperature and humidity also play important roles. Dry environments particularly inhibit microbial growth.

Application design will begin with establishing what the optimum degrade curve should look like for a given application. For many applications, the need for long lifetimes of use is not an issue. Consider how many plastic products are used and then discarded within weeks or days or even hours after they are purchased. A large amount of packaging is discarded once the article arrives home. In such cases the packaging has to survive only the periods of distribution and shelf life. Natural antimicrobial additives—sorbic acid is one—can then be used as needed. In any given application, a carefully determined amount of natural preservative will allow controlled degradation, and formulations can be developed to achieve optimum degrade curves.

Bioplastics, through their properties, have to meet the multiple challenges of remaining stable during storage and use, performing intended functions, and then biodegrading at the appropriate time under the planned disposal conditions.

Cost

The technical questions connected with developing adequate physical properties lie in the area of materials science. Scientists and engineers have to optimize properties for given applications through variations of composition and processing, to discover more and more useful new products, but there are development costs associated with the search for commercially useful formulations and processes.

In addition, commercially available biopolymers are typically more expensive than synthetic resins, often significantly so. Currently only starch competes with synthetic resins in terms of cost.

It is too early to tell how much the costs of raw materials might be brought down by a growing industry and the resulting incentive for increased production of raw materials.

There are processing costs as well, and it is likewise too early to tell how much costs can be brought down by using large-scale manufacturing, or by using novel processing methods that minimize energy use and waste by-products. The nascent bioplastics industry is also at a competitive disadvantage relative to the mature conventional plastics industry, because it needs investment capital. There are already useful patents awaiting commercialization.

All these factors place economic hurdles in the path of the commercialization of bioplastics. Successful commercialization will be a matter of "picking oneself up by the bootstraps." To develop extensive markets one needs low-cost products. To achieve low-cost products, one needs large-scale production and extensive markets.

In considering cost it is usual not to include such factors as the depletion of natural resources, the costs of guaranteeing crude oil supplies, or the environmental burden of waste management. A total cost accounting, a *green accounting,* would include those factors. There is nothing close to a consensus on how a total accounting should be carried out, but the approaching exhaustion of fossil fuels and growing waste management problems make all hidden costs more and more relevant, and resolution of the accounting question is becoming increasingly important. A United States National Research Council study panel reported in 1999 that the United States now lags behind other countries in developing a national accounting method that would link economic growth with the consumption of natural resources.

Even without a total green accounting there is already a trend toward agricultural raw materials becoming cheaper relative to oil. Although there have been short-term fluctuations in costs over the years, it was nevertheless possible to buy four times as much wheat with one ton of oil in 1990 as in 1967.

If the required physical properties can be achieved and costs adequately managed, then all of the advantages of bioplastics can be realized. How the factors favoring and impeding growth are intertwined with societal concern for resource conservation and environmental preservation is described in the final chapter.

9

Prospects for the Future

Raw Materials

What role will bioplastics play in the future? At present, no one knows. Their future will depend on several interrelated factors, the first of which concerns the availability of raw materials. Currently, the use of renewable biomass for bioplastics is negligible, even though their raw materials are not only in abundant supply but also are being renewed constantly. We will only begin to make full use of these resources if there are economic incentives for their producers, such as the demand brought about by a growing market. Increased demand will depend on developing useful materials that can succeed in the marketplace. If there is an economic benefit to the producers of the raw materials, the availability of the raw materials will grow.

Because some of the raw materials are food crops, there may be those who wonder about using them as feedstocks for a bioplastics industry. People might worry that producing bioplastics could be using resources needed for food. But that assumes food crops production is at its capacity, and present production is not a good indication of production potential. The current situation in the major agricultural producing countries is that production potential is larger than present production because government policies in the United States, Canada, Western Europe, and Japan have restricted agricultural production in order to keep it within market demand. By one estimate, more than one-third of the total land used today for agriculture could be taken out of production by the year 2030 on account of increasing productivity.

For example, in the United States and Canada, soybean growers are very efficient, with modern soybean production methods and highly advanced processing technologies. Production could be increased were it not that existing soy protein markets in animal feeds are saturated. Additional production would simply depress prices. New uses for soy protein and soy oil would allow production increases.

A bioplastics industry would provide new markets for low-valued biomass commodities, both agricultural and marine. Underutilized land would be reactivated and languishing rural areas revitalized.

In the long term—fifty to a hundred years in the future—even the sustainability of agriculture will be challenged by world population growth. During the twentieth century the world's population tripled, and there are now six billion people on Earth, with expectations of billions more in only a few decades. In the United States and elsewhere, prime farmland is being lost to development, topsoil is being depleted through erosion, and the aquifer is being drained. More land will eventually have to be set aside for biomass energy production. Population growth will in time result in allocations of land for food, land for energy, and land for nonfood uses. Ocean resources will necessarily take on greater importance. Discussions of long-term land use and allocation have not yet begun in earnest, and perhaps they will not begin until the primary question of population growth is confronted squarely. Only then will it be possible to face important long-term global issues like land use, water supplies, climate change, and pollution.

Currently, however, and for the intermediate future, land and other agricultural resources are sufficient to meet demands for food, animal feed, and fiber, and still produce raw materials for industrial products; we have a surplus of farm commodities. The environmental consequences of increasing biomass production for bioplastics and other nonfood uses will be positive, even when the environmental cost of increased agricultural production is factored in. In any case, the inevitable exhaustion of fossil resources will make increased production necessary, and the only question is how well prepared we will be for that necessity.

Feedstocks can be diversified. Research has shown that a variety of commodity plant oils and animal fats can be used as carbon sources for commercial fermentation processes. For example, rapeseed oil methyl ester is already being produced in large amounts for use as a biodiesel fuel in automobiles. Research shows that it also can provide the sole carbon source for the production of biodegradable polyhydroxyalkanoate polyesters. Olive oil, sunflower oil, soybean oil, and coconut oil also have been studied for use as fermentation feedstocks, as well as butter oil and animal fats.

There are even projects under way aiming to produce polyhydroxyalkanoates using food industry wastes as a carbon source, and municipal sludge to provide a mixed culture of microorganisms.

In the future, through **biotechnology**, the genetic manipulation of microorganisms and plants will give rise to new types of improved biopolymers for nonfood uses. Biotechnology aimed at producing bioplastics has already begun. In 1987 the genes of the bacterium *Alcaligenes eutrophus*—now named *Ralstonia eutropha*—which is capable of producing the polyester PHB, were cloned and inserted into the bacterium *Escherichia coli*, a bacterium that normally does not produce PHB. The modified *E. coli*, with its new genes, began producing PHB as expected. Modifying *plants* to produce polyesters instead of starch is not so simple, but one can envision a day when potatoes, for instance, could be bioengineered to produce plastics. Although these *transgenic* approaches are intended for small-scale production in highly specialized applications, it may one day become possible to produce some materials on larger scales. For example, increases in forestation rates through genetic modifications may provide a more rapidly renewable source of cellulose.

Biotechnology will also eventually make it practical to degrade renewable biomass to generate large amounts of the fermentable sugars used in the production of polymers like poly(lactic acid) and microbial polyesters. Current major feedstocks for fermentable sugars are planted crops, like corn, wheat, sorghum, potatoes, sugar beets, and sugarcane; and agricultural by-products such as potato-processing residue, sugar beet and sugarcane molasses, and apple pomace. But woody plant parts—lignocellulosics—are potentially vast sources of fermentable sugars that could increase the supply by orders of magnitude. Nonfood agricultural crops could be grown on land unsuitable, or only marginally suitable, for food crops, particularly perennial grasses like switchgrass, rye, and napiergrass. Forest crops—especially short-rotation woods like hybrid poplar, sycamore, and silver maple—could add to the supply of feedstocks. The crops ultimately selected for such use will be chosen for their high yield per acre and low-maintenance requirements.

Waste from agricultural, industrial, and municipal waste streams provide another potential source of feedstocks. In the United States 500 billion pounds of lignocellulosic waste are generated each year, including 200 billion pounds of corn stover, 120 billion pounds generated in paper mills, and 75 billion pounds of urban tree residue. The remainder includes other crop residues, wastepaper, and other wastes.

Developing technologies for converting lignocellulosic material to fermentable sugars is an active area of research—no commercially viable technology yet exists. It may be simply a matter of learning more about the chemistry of lignocellulosics. What is known is that direct acid hydrolysis destroys much of the desired glucose. It is also known that enzymatic digestion, with the enzyme cellulase, is slow—and the enzyme is expensive. One major current effort is to develop an efficient *pretreatment* of the lignocellulosic material so as to make an enzymatic hydrolysis effective.

Finding additional commercial uses for *lignin* would also increase the competitiveness of *any* conversion technology; the value of the lignin would increase and its disposal costs would be avoided. Lignin chemistry has become recognized as a high research priority in government, industrial, and academic laboratories. Scientists at IBM, for example, have investigated the use of epoxidized lignin to make thermosetting resins for printed wiring boards.

As more and more attention is given to how biomass can be exploited as a renewable resource, progress in converting biomass to chemical feedstocks—as well as to fuel—will become more likely.

For now, nature has provided us with an abundant supply of annually renewable resources to be used as raw materials, and that readily available supply can be exploited.

Markets

In developing useful bioplastics products, some markets are more accessible than others. Companies have already developed commercial products made from bioplastics (Chapter 7), and in many cases those markets are being developed successfully. Some of the active market areas for bioplastics are

- garbage bags and compost bags
- trash bin liners
- single use disposable packaging film
- single use or limited use disposable packaging materials, such as recyclable and compostable bottles
- loose fill packaging
- rigid foam packaging
- food service items for institutional use, the fast food industry, and the general public
- agricultural products, such as mulch films, silage bags, temporary covers for storing grain, netting to protect crops from birds, binders and twine, irrigation products, and containers for plant nurseries
- products for marine applications, in all of the above categories, intended for disposal at sea
- fishing netting and fishing lines
- personal care products such as combs and razor handles
- sports items such as golf tees
- pet products such as pet toys
- coatings, especially for paper products

What markets can be developed in the future will depend on further advances in technology. Food packaging has especially been targeted and is likely to be one of the most quickly developed markets for bioplastics. In the area of specialized biomedical applications, the initial target markets that have already been developed have included wound closure products, drug delivery systems, and orthopedic repair products. In biomedical applications, high costs are less likely to impede market growth.

In commercializing bioplastics and developing markets for them, the "bottom line" will be costs relative to performance, but the cost that can be tolerated will depend on the application. For example, in Japan, where farms are relatively small, polyethylene mulch films are manually removed after the growth season and incinerated. A biodegradable product that is more expensive than nondegradable mulch film is still acceptable to farmers when there are sufficient savings in labor costs and disposal costs.

Technological Advances

A bioplastic has to have properties adequate for its intended application. The commercial ventures already under way in the United States, Canada, Europe, and Japan indicate there is confidence that technological advances are possible. The key to solving technical problems is often simply knowing what the problems are.

Developing useful bioplastics lies in the area of applied science, particularly the chemistry, biology, and engineering of materials science. The large variety of potential compositions gives cause for optimism. The large body of technical knowledge for conventional plastics is also an advantage; much of that technology can be applied to bioplastics. For that reason, it is clear that as the technology of bioplastics advances, the usual terms and definitions of the plastics industry will be employed, and standard industry tests will be applied to demonstrate the usefulness of the materials.

Conversion of raw materials into marketable end products will require a variety of processes—chemical, biological, and engineering. Some of the required technologies have been tested on a large enough scale to indicate they are commercially practical. Others are still in the evaluative stage. In the future one can envision large *bio*refineries becoming more and more developed until they begin to match the *oil* refineries of today.

Advances in *biotechnology*, including genetic engineering, will play a role in the development of new and improved feedstocks. Novel genetically engineered microorganisms will likely play a role in the large-scale conversion of lignocellulosic materials into fermentable sugars. Enzymatic processes will likely be developed for producing high-purity lignin needed, for example, in thermosetting-resin applications. Microbial agents will also be designed to optimize biodegradation mechanisms for specific disposal environments.

The use of enzymes to *catalyze* bulk polymerization is being developed. Enzymes are already in widespread use in the food, pharmaceutical, detergent, and textile industries. Enzyme technology is one example of "green chemistry," where the chemical process takes place in aqueous or nonsolvent environments, at ambient temperature and pressure, and without the need for metal catalysts.

In developing the technology, judgment will have to be exercised on the question of how "green" the technology is to become. For example, introducing petroleum-derived components into formulations may lower costs and perhaps improve physical properties, at least in the short term. Examples are the use of poly(vinyl alcohol), poly(glycolic acid), polycaprolactone, or poly(ethylene glycol) in formulations. Using them represents only a partial compromise because, although derived from petroleum, they are biodegradable to the point of complete mineralization. Also, there are many different surface coatings used with conventional plastics to improve physical properties. Some are more environmentally benign than others. Environmental issues will play a prominent role in decision making.

At the present time there is no "magic bullet" in the field of bioplastics. Perhaps the magic bullet will be a starch-based bioplastic, formulated to take advantage of the low cost and thermoplastic nature of starch, but containing additional components—polymeric or otherwise—that give the final composition excellent physical properties suitable for a wide range of applications. Perhaps the magic bullet will be a product of fermentation technology, made attractively inexpensive with technological advances that allow large amounts of fermentable sugars to be produced inexpensively from agricultural crops or waste. Perhaps it will come from some unanticipated direction. Or there may just be incremental improvements in several technologies, leading to an increasingly diverse range of useful materials, and a proliferation of products in the marketplace.

Predicting technological advancement is not easy; there will always be optimists and pessimists. The pessimists have sometimes been dramatically wrong, probably because people are so ingenious and resourceful. One example is worth recounting. Many people are familiar with the medical technique of magnetic resonance imaging (MRI), now widely used in hospitals. MRI requires a magnet that not only must produce a very large magnetic field, but also one that is "smooth," or homogeneous, within a defined region of space. It also must not vary in time; it has to be very stable. A few decades ago, scientists were contemplating the prospects of being able to produce very large field strengths and still maintain homogeneity and stability. Some members of the scientific community were pessimistic because the technical difficulties were very great. Some flatly said that stable, homogeneous magnetic fields could not be made any larger than they already were. It turned out that the pessimists were proved wrong, and in an embarrassingly short period of time. There was a need, the need spurred efforts, and the efforts were successful.

It remains to be seen what level of motivation exists for the development of bioplastics. The question to be asked is, how strong is the need for moving toward the use of bioplastics?

Environmental Concern

Much of the desire to find a place for bioplastics in this world of plastics will stem from a concern for the environment, and interest in the development of bioplastics will grow largely to the extent that there is real interest in and concern over the environment. The picture that some environmentalists paint about the depletion of fossil fuels, the accumulation of waste, and other acute environmental problems is not a pretty one. No one can say for sure how the future will play out on these crucial issues, but the price of being underconcerned could be extraordinarily great.

A "cradle-to-grave" green technology for plastics starts with renewable resources and ends with biodegradable products that can be returned to natural biogeochemical cycles. ("Cradle-to-cradle" might be a more apt expression.) For some people, totally green-technology is the only acceptable solution, while others are willing to compromise. In the end, how "green" the emerging technology becomes will reflect the concern of the community. Even small steps will reduce our use of nonrenewable resources and ease the production of waste. So no matter what the ultimate goal is, improvements can be made now.

Environmental concern is already being expressed through government agencies, business and industry, scholars and academics, and the general public.

The Role of Government

Restrictive legislation on the use of plastics, particularly aimed at plastic packaging, has already begun at the local, state, federal, and international levels. The United States Plastics Pollution Research and Control Act of 1987 (Public Law 100–220) and Annex V of the international *MARPOL Convention* ("The International Convention for the Prevention of Pollution From Ships") prohibit the disposal of plastics at sea. For the United States Navy that prohibition went into effect in January 1994. The Navy, in order to comply, has promoted the development of biodegradable plastics for eating utensils, drink cups and lids, food trays, and trash bags. In 1990 the United States Congress amended the 1987 legislation with Public Law 100–556, which requires that plastic ring-carriers for bottles and cans be made of degradable material. Trash dumped into the sea and plastic ring-carriers that become litter pose threats to living creatures and to the environment in general, whether they are degradable or not. The positive impact of the legislation lies in limiting the *lifetime* of the discarded materials and, thereby, diminishing their adverse effects.

Legislation will undoubtedly increase in the future, and new legislation will likely contain restrictions aimed at materials that are neither recyclable nor biodegradable. Packaging plastics may ultimately be required to be compatible with composting. Environmental tax legislation will probably increase on products deemed harmful to the environment. Labeling legislation may lead to an "ecolabel," based on a product's raw material usage, energy consumption, emissions from manufacture and use, and waste disposal impact.

In Europe, Germany has pioneered legislation for packaging waste. In 1991 Germany made industry and trade responsible for used packaging through the "Green Dot" (*ein Grüner Punkt*) system. Packagers are now obliged either to collect their own waste for recycling or to contract with the company formed to oversee the system, Duales System Deutschland (DSD), for collection and recycling. A small green dot labels products of contracting producers. In 1999 11 billion pounds of used commercial packaging were recovered, including more than a billion pounds of plastics.

In Japan, organic recycling was as extensive as in any other country before the Industrial Revolution, and Japan is moving back toward that position, partly because the large use of incineration has led to increasing levels of dioxins in the environment. On-site burning of used polyethylene agricultural covers was banned in 1997, and the approximately 10 million pounds of it used each year are now collected for incineration at regional facilities where emissions can be better controlled. In order to eliminate the need for incineration entirely, farmers cooperatives are testing degradable mulch films as a replacement for polyethylene. The large fast-food service industry in Japan is looking for biodegradable, microwavable trays and other serviceware, which may soon have to pass ISO compostability tests. Concern is growing in other countries of Asia as well, including Korea and Taiwan.

Now that industry standards for compostable plastics have been developed, legislation requiring compostability will be easier to formulate and may increase for more and more products. Legislation at all levels of government will no doubt continue to change the way in which waste disposal is viewed. Bioplastics provide an environmentally sound response to the concerns that motivate such legislation.

Government agencies will also play a role through their procurement policies. In the United States, the Department of Defense, Department of Agriculture, Department of the Interior, Department of Energy, and other agencies are major purchasing powers. The Defense Logistics Agency, within the Department of Defense, is the largest procurement office in the United States government, having oversight of $6.4 billion in nonweapons defense spending. Federal agencies are likely to become influential in promoting the development and manufacture of environment friendly products, especially food service items, waste disposal materials, and personal use articles. Executive Order 13101, "Greening the Government," mandated government agencies to buy more recycled products. Executive Order 13134, "Developing and Promoting Biobased Products and Bioenergy," extended the scope of the previous order. An interagency working group responsible for implementing it has defined *biobased products* as "commercial or industrial products (other than food or feed) that utilize biological products or renewable domestic agricultural (plant, animal, and marine or forestry) materials." One of the first items on its agenda was to develop a nationally recognized biobased cutlery standard that could be used in federal solicitations.

These executive branch actions were followed by the Biomass Research and Development Act of 2000, passed by the United States Congress. It established the Biomass Research and Development Board to coordinate federal programs in promoting the use of biobased industrial products. It also called for a Biomass Research and Development initiative to support research that promotes biobased products, including "research on process technology aimed at overcoming the recalcitrance of cellulosic biomass, and research on diversifying the range of products that could be made from biomass."

State, county, and local governments are similarly able to promote environment friendly products through their purchasing guidelines for law enforcement agencies, colleges and universities, school systems, hospitals, prison systems, and other institutions in their jurisdiction. Individual institutions, public and private, have already begun to initiate their own purchasing guidelines without waiting for mandates from above.

Through restrictive legislation and procurement policies, government bodies can and will stimulate and encourage industries that use renewable resources in the manufacture of environmentally degradable products.

The Role of the Private Sector

Stimulus and encouragement are all that government agencies can provide. The major driving force for moving toward sustainability may prove to be the private sector, and many business leaders feel that industry, not government, will play the most important role in environmental preservation and resource conservation. Companies are getting involved more and more in environmental stewardship, and their approach to stewardship has been evolving. At first, the main interest was to be compliance-oriented, aimed at dealing with government-imposed environmental requirements, compliance costs, liabilities, and violations—well after design and production decisions had been made.

More and more companies have expanded their approach by integrating environmental stewardship into their planning from the very beginning. For example, for some time there has been a growing trend from "end-of-the-pipeline" recycling to waste reduction. Reducing energy or raw material saves money at the same time that it creates environmental value. More broadly, design and production decisions, while still aimed at economic advantage, are now having environmental stewardship concerns incorporated into them from the very beginning, including waste reduction savings and pollution prevention.

Responsible Care, a worldwide industry program of environmental stewardship, has spread to forty-two countries that together account for 85 percent of world chemical production. In the United States there are now more than 300 community advisory panels, and since 1988 the release of toxic chemicals to air, land, and water has dropped 50 percent. The American Chemical Society has endorsed the Responsible Care initiative and has been promoting awareness of it through its activities.

In industry, environmental performance is more and more widely viewed as a component of competitive advantage. Companies may be motivated partly by altruism, but their action is also dictated by cold business logic. Environmental concern is viewed as something that will eventually be required of all companies, and businesses will thrive—or even survive—to the extent that they are prepared. In the future, environmental factors will likely have even more impact on business decisions, particularly in overall strategy and product design.

Attention, at least among the largest companies, has begun to move beyond ecoefficiency and waste reduction to genuine **sustainable development**, defined by the United Nations' World Commission on Environment and Development as *"development that meets the needs of the present without compromising the ability of future generations to meet their own needs."* Sustainable development, as the only acceptable kind of development, was the theme of the United Nations Conference on Environment and Development in Rio de Janeiro in June 1992. Sustainable development includes the more traditional elements of energy efficiency, pollution prevention, and environmental auditing, but it also includes the newer and more demanding areas of design for environment and life-cycle assessment. The term *sustainable development* may still be met in some business quarters with a blank stare, but a growing number of companies are integrating the concept into their business operations for purposes of self-interest. Whether it was ever true that the goals of industry diverge from the goals of preserving and enhancing the environment, today there are many signs that at least forward-looking corporations see a need to pay serious attention to the relationship between industry and the environment.

The International Standards Organization (ISO) has been defining standards of quality in manufacturing for some time, and it has become familiar to a growing number of businesses through its ISO 9000 process of registration. Its current Technical Committee 207 has been working since 1993 to develop the 14000 Series Standard, entitled "Standardizations in the Field of Environmental Management." The committee's task is to study life-cycle assessment programs, develop product standards that incorporate environmental aspects, generate environmental auditing procedures, and develop meaningful product-labeling programs, beginning with definitions of terms like *recyclable, compostable,* and *biodegradable.* Its overall aim is to foster environmentally sustainable industrial development.

A new expression of environmental concern within the business and academic communities has been the development of the discipline of **industrial ecology**, which has been defined as the *science of sustainability*. Industrial ecology applies a systems approach to the use of materials, energy, and products, and encompasses a wide range of topics for analysis. The topics include the environmental impacts both "upstream and downstream" from manufacturing, following materials and energy from their source, through their conversion to products, to their final integration into natural biogeochemical cycles.

Discussions of the concepts of industrial ecology go back at least to the 1970s, but by the early 1990s the United States National Academy of Sciences and National Academy of Engineering were holding colloquia and workshops on the topic. Now the discipline has its own academic titles, graduate students, federal grants, textbooks, and research journal (*Journal of Industrial Ecology*). In 1998 the first Gordon Research Conference on industrial ecology was held in New Hampshire.

Environmental concerns over resource use, product manufacture and use, and waste disposal are receiving increasing attention on all fronts—from governments, industry, and academia—and the increase in attention will no doubt continue.

Paradigm Shift

Most of all, what is needed for bioplastics to find a place in the current Age of Plastics is a paradigm shift. We have grown accustomed to having a wide variety of useful plastic materials that are attractive, long lasting, and inexpensive.

On the other hand, we are coming to realize, in retrospect, that we may have had too much of a good thing, and have given too little thought about the effect their continually increasing use has on the future. They drain irreplaceable resources and, once manufactured, the sturdiness that has been imparted to them makes them persist long after they have served their useful purpose, causing them to be relegated to mausoleums of discarded waste. There is something downright silly about wrapping a sandwich in a package that will last fifty years.

In ignoring nature's way of building strong materials, we have, for many applications, overengineered our plastics for stability, with little consideration of their recyclability or ultimate fate, and have ended up transforming irreplaceable resources into mountains of waste.

There is another way. Plants and animals have been producing strong, pliant materials for eons. Plants produce these materials by using energy from the sun, harnessed through photosynthesis. Animals produce them by using the energy stored in plants. When the plants and animals die, the materials degrade naturally, so that they can be recycled. Atoms are continually being rearranged in this chemistry of nature, through cycles of life and renewal. We can take nature's building materials and use them for our purposes, without taking them out of nature's cycles. We can be borrowers, not consumers, so that the process can continue indefinitely.

Environment-related paradigm shifts have occurred before. Many solid-waste-disposal managers at first considered recycling to be needless interference. A very common viewpoint was that sorting by homeowners over any sufficiently practical period of time was simply infeasible. Incineration and landfills were considered the *only* realistic alternatives. In time, however, that view has changed.

If there is no paradigm shift, bioplastics will only find their way into easy "niche" markets. But then, we will still be left with the question of what to do with the tens of billions of pounds of spent plastics generated annually. And when we run out of petroleum, we will not have mastered the technology of manufacturing plastics from nature's polymers. We will have to look for a makeshift solution, possibly involving the direct synthesis of chemical feedstocks of the type now being used, and requiring large amounts of energy that only nuclear sources can provide. Is that the future we want? It could be the future we get.

There are many signs, however, that society is becoming more and more committed to the concepts of resource conservation and environmental preservation. We now see environmental questions not merely as technical matters to be left to the experts, but as questions bearing on who we are and what legacy we wish to leave.

Events are regularly taking place that reveal a growing interest in the environment and an accelerating pace of activity.

• The field of biodegradable plastics from renewable resources now has its own scientific societies, like the United States–based BioEnvironmental Polymer Society, the European Society of Biodegradable Polymers and Plastics, and the Biodegradable Plastics Society of Japan. National and international conferences devoted to the subject are held regularly, and specialized scientific journals now exist, like *The Journal of Polymers and the Environment* and *Biomacromolecules.*

• The scientific program for the Fifth Chemical Congress of North America, held in Cancun, Mexico, in November 1997, featured a symposium on "Natural Polymers as Advanced Materials: Polymer Degradability and Performance." The talks included "Water-Soluble Polymers from Natural Raw Materials: Why the Driving Force and What Are the Chances of Success?" presented by G. Swift, and "Processing and Characterization of Biodegradable Products Based on Starch and Cellulose Fibers," presented by W. J. Bergthaller.

• Chrysler Corporation announced its Composite Concept Vehicle (CCV) at the 1997 Frankfurt International Auto Show. It has a body made of four plastic panels mounted on a steel chassis. The plastic panels are 85 percent poly(ethylene terephthalate) (PET). The plastic can include up to 20 percent recycled resin, can itself be recycled, and needs no paint. Although PET is a petroleum-based synthetic resin and not a bioplastic, the introduction of the concept of a recyclable plastic automobile body by a major automobile manufacturer shows how far we have come. And it is not entirely far-fetched to imagine taking the next step, whereby a durable bioplastic composite is found for the same purpose, recalling Henry Ford's soybean automobile body and perhaps leading, after all, to a new slang expression, "—or I'll eat my car!"

• The joint Cargill–Dow Polymers venture, which aims to develop and market poly(lactic acid) resins, was announced in 1998. Global capacity is planned to increase to about a billion pounds per year within the next decade. The venture, which joins the forces of an agricultural company and a chemical company, is not the only one of its kind, but may signal a growing trend in business arrangements. In the future, the *biological* plants of agriculture and the *manufacturing* plants of the chemical industry may simply merge in one large advantageous symbiosis.

• In a 1998 panel discussion sponsored by the American Chemical Society (ACS) of what the next twenty-five years hold in store for the chemical industry, moderator Dr. Paul S. Anderson, a former president of the ACS and senior vice president for chemical and physical sciences at DuPont Merck Pharmaceutical Company, predicted that plants "will actually become the main source of oil and plastics . . . the rudimentary technology for that already exists today." A shift from hydrocarbon feedstocks to feedstocks based on plants is already beginning to take shape and, ultimately, will become a necessity. As one analyst put it, "We're moving to a carbohydrate-based economy."

• The American Chemical Society also sponsored a major symposium, entitled "Polymers from Renewable Resources," at its August 1998 National Meeting held in Boston, Massachusetts. Among the more than fifty research reports presented were the following: "Structure and Function of Polyhydroxyalkanoate Inclusion Bodies," by L.J.R. Foster, R. C. Fuller, and R. W. Lenz; "Extrusion Processing of Soy Protein–Based Foams," by P. M. Mungara, J. Zhang, and J.-L. Jane; "Biodegradable Films from Selectively Modified Feather Keratin Dispersions," by P.M.M. Schrooyen, P. J. Dijkstra, and J. Feijen; "Preparation, Solubilization and Biodegradation of Crosslinked Gelatin," by R. D. Patil, J. E. Mark, P. Dalev, E. Vassileva, and S. Fakirov; and "Role of Glycerol in Plasticized Starches," by S.H.D. Hulleman, F.H.P. Jansen, F. C. Ruhnau, J.J.G. van Soest, H. Feil, and J.F.G. Vliegenthart.

• The International Union of Materials Research Societies presented an innovation award in 1998 to, among others, Richard P. Wool, professor of chemical engineering at the University of Delaware at Newark, whose team developed the high-strength, low-cost thermosetting resins described in Chapter 7 (fig. 7.11).

• In 1999 the United States Department of Energy, Office of Industrial Technology, continued its support of the technology agenda that was announced in its "Technology Vision 2020" report. Project topics included developing inexpensive cellulose for plastics.

• At the 1999 meeting of the Bio/Environmentally Degradable Polymer Society—precursor of the BioEnvironmental Polymer Society—sixty-eight research reports were presented, including "Biodegradable Composite from Wheat Straw and Proteins," by J. Hu, X. Mo, X. S. Sun, and J. A. Ratto; "Compostability of Starch-$CaCO_3$ Disposable Packaging," by V. T. Breslin; "Bioplastics and the Waste Management Crisis in Japan," by D. Kitch; "Manufacture of Glass Fiber Reinforced Soy-Protein Polymer," by F. Liang, Y. Wang, and S. Sun; and "Microbial Synthesis of Poly(3-hydroxybutyrate-*co*-3-hydroxyalkanoates) from Renewable Resources by Recombinant Bacteria," by Y. Doi and H. Matsusaki.

• At the March 2000 National Meeting of the American Chemical Society in San Francisco, California, symposia were held on "Biobased Processing to Chemicals" and "Frontiers for Polymer Science in the 21st Century." Research reports included "Biobased Economy of the 21st Century," by R.W.F. Hardy; "Toward the Commercialization of Biodegradable Plastics in Japan," by M. Matsuo, K. Fukuda, and N. Kawashima; "Biobased Materials for Sustainable Growth," by R. R. Dorsch; and "Poly(lactic acid): Performance Materials from Renewable Resources," by M. N. Mang.

• In the United States, the percentage of municipal solid waste that is recycled, including that composted, increased from 16 percent to 28 percent during the 1990s. In the same period the percentage disposed of in landfills dropped from 67 percent to 55 percent, a result few had believed possible.

• The United States Congress passed the Biomass Research and Development Act of 2000, aimed at supporting research on biobased industrial products and promoting their use.

• In 2000 Japan passed the Basic Law for the Creation of a Sustainable Society, encouraging the use of renewable raw materials and the manufacture of biobased products. Also in 2000, the Japanese Organic Recycling Association (JORA) was founded in Tokyo. Its purpose is to promote organic recycling in Japan, so as not to continue the incineration of more than 90 percent of Japan's waste.

• In December 2000 an International Scientific Workshop on Biodegradable Polymers and Plastics was held in Hawaii. Over a hundred scientific papers were presented by scientists from all parts of the world, including the United States, Taiwan, Sweden, the Netherlands, New Zealand, Korea, Japan, Italy, Germany, France, Finland, and England. Reports included "Biodegradable Nanocomposite Food Packaging," by J. J. de Vlieger, S. Fischer, L. Batenburg, and H. Fischer; "Production of Poly(3-hydroxybutyrate) by Activated Sludge Treating Domestic Sewage," by H. Satoh, H. Takabatake, A.S.M. Chua, T. Mino, and T. Matsuo; "Biodegradation of Natural and Synthetic Polyesters under Anaerobic Conditions," by R.-J. Müller, D.-M. Abou-Zeid, and W.-D. Deckwer; "Rheological and Barrier Properties of Edible Films in Relation to Microstructure," by M. Stading and M. Anker; "Plasticized Wheat Starch Based Biodegradable Blends and Composites," by L. Avérous, O. Martin, and L. Moro; "Bonding Fiberboards Using Components Coming from the Fiber—A New Look at an Old Idea," by R. Rowell, S. Lange, and M. Davis; "Use and Effects of Agricultural Fibers and Fillers in Baked Starch-Based Foam Composites," by G.A.R. Nobes, W. J. Orts, and G. M. Glenn; "Composites from Soy-Based Resins Reinforced with Natural Fibers," by Z. S. Petrovic, W. Zhang, I. Javni, and A. Guo; and "Cellulose Nanocrystals: Surface Modifications for Use in Stimuli Responsive Materials and Composites," by W. T. Winter, A. J. Stipanovic, D. Bhattacharya, M. Grunert, and S. Zhang. The reports demonstrate the expanding scope of scientific investigation and growing global interest in biodegradable plastics made from renewable resources.

9. Prospects for the Future

By 2001, government agencies throughout the world were supporting research on biobased materials, including *green chemistry*—the development of materials and processes that are environmentally friendly. Also targeted are bioprocessing technologies to increase energy efficiency, and engineering technologies for converting waste biomass into useful feedstocks. In the United States, the Department of Agriculture continued research aimed at the development of new biobased products, so as to provide additional markets for agricultural crops. The National Science Foundation and Environmental Protection Agency renewed their interagency partnership in the "Technology for a Sustainable Environment" program.

During 2001, established bioplastics companies expanded and new commercial ventures began, with the aim of bringing to the public a larger and more diverse array of products.

If society is indeed becoming more and more committed to resource conservation, environmental preservation, and sustainable technologies, bioplastics will find their place in this Age of Plastics.

Appendix

Make Your Own

Preparation of Cast-Film Bioplastics

Clearly, making bioplastics takes some processing, but there are examples that can be made easily and with little cost. Instructions are given here for anyone interested in exploring bioplastics firsthand. The preparation method involves the casting of films by evaporation of solvent, where the solvent is water. The method could be considered an illustration of small-scale cast-film technology. The resulting materials may not be as strong as needed for some purposes—like heavy-duty trash bags—but they provide direct experience with bioplastics—biodegradable plastics made from renewable raw materials.

Preparing bioplastics on the small scale described here provides an easy introduction to the practice of science. The formulations are simple, and the raw materials are available on the shelves of supermarkets and pharmacies. The projects are easily adapted for instructional use.

The preparations can be varied in any number of ways, limited only by one's imagination. There are so many conceivable compositions and applications that there is even opportunity for novel discoveries. A few years ago, two students in the state of Virginia accidentally produced a gelatin-based material that attracted the interest of a pharmaceutical company as a potentially useful drug-encapsulating material. The students and their teacher shared a $100,000 fee paid them by the company!

The *safety considerations* in the following experiments are not extensive. Bioplastics are prepared from formulations similar to those used to make gelatin desserts. Observe the following safety rules:

(1) Handle hot liquids with care—the mixtures are heated to nearly the boiling point.

(2) Protect your eyes from splashes and other mishaps. Many states require all laboratory workers to wear proper safety glasses for eye protection.

(3) Wash your hands after every laboratory session.

(4) Do not put the final products in your mouth for any reason. Leave it to the experts to determine ingestibility.

(5) Young children should not experiment without adult supervision.

Supplies

The supplies that are needed are commonly available and are inexpensive. Recalling the general formula for a bioplastic, you see that you will need at least one biopolymer and at least one plasticizer. Three useful bioplastic components are readily available:

Gelatin. Gelatin is an agricultural protein derived from animals. Unflavored gelatin is widely available as a colorless or pale yellow granular solid that is sold in food markets.

Starch. Starch is an agricultural polysaccharide derived from plants. It is an important feedstock for bioplastics on account of its large-scale availability and low cost. Cornstarch is widely available as a fine white powder in food markets.

Glycerol. Glycerol, also called glycerin, makes a very useful plasticizer. Glycerol is produced by the fermentation of sugar, or from vegetable and animal oils and fats, as a by-product in the manufacture of soaps and fatty acids. Glycerol is often available in drugstores. It is a liquid at room temperature. (The water used in the following recipes also acts as a plasticizer.)

Teachers or students in high school or college will have access to a much wider range of materials through chemical supply houses. *Agar* and *sorbitol* are two useful materials, in addition to the three already described. Agar is a marine polysaccharide derived from red seaweed. Sorbitol, like glycerol, acts as a plasticizer.

The mechanism by which glycerol and other small molecules, like sorbitol, increase the mechanical plasticity of cast films and sheets is still not completely understood. In recent research on starch-sorbitol films cast from water, the local mobility of the starch molecules has actually been found to

Continued on next page

Continued from previous page

be *reduced* by the addition of sorbitol (when it is present at less than 27 percent)—the sorbitol is acting as an *antiplasticizer*. Presumably the sorbitol forms cross-links with the starch, and the cross-links make the starch molecules less mobile. The observed increase in mechanical plasticity is pictured as arising from the formation of clusters of sorbitol molecules. Glycerol may act in a similar fashion—by forming cross-links with starch molecules the starch mobility is decreased, but clusters of glycerol molecules increase film plasticity.

Glycerol and sorbitol might better be referred to simply as "processing aids" rather than plasticizers, but here—in keeping with current usage—they are called plasticizers, where the word is meant in the general, descriptive sense of agents that increase overall mechanical plasticity, and not in the more technical sense of increasing polymer mobility.

Although you need at least one biopolymer and one plasticizer, you can use different amounts of each, and when you have access to different kinds of biopolymers or plasticizers, you can try different combinations as well. Each component will contribute to the overall properties of the plastic in its own way. To make a plastic with particular desirable properties may take a very special mix.

For example, the more plasticizer present, the more flexible the plastic becomes, but too much plasticizer makes it tacky. The less plasticizer the stronger the plastic, but it will be less flexible also. Trade-offs play an important role in deciding what the best formulation is for a particular application. Glycerol is an effective plasticizer and inexpensive, and it tends to make the resulting plastic flexible even at the very low temperatures of a freezer, as might be required for a freezer wrap. On the other hand, too much of it makes the plastic curl up in a microwave oven and turn to gum. Even more important, glycerol tends to lose its effectiveness as a plasticizing agent over time, leading to a slow increase in brittleness. Sorbitol produces better resistance to microwave radiation and does not lose its effectiveness as a plasticizer as quickly as glycerol. On the other hand, it typically leads to brittleness at freezer temperatures, and if there is too much of it, chalking tends to occur whereby the sorbitol crystallizes out, producing a white, spotty appearance.

Agar, either by itself or in blends with other biopolymers, appears to impart favorable properties to plastic sheets. In plastics containing agar and glycerol, the effectiveness of the glycerol lasts longer, because the agar seems to slow down the increase in brittleness. Agar also seems to improve resistance to microwave radiation, and it improves clarity in sorbitol formulations. Agar is more expensive than starch, which limits its large-scale use. But as applications for agar plastics are developed and the markets for them grow, a greater inducement for increased commercial production may lead to reduced cost. One reason for agar's high cost is the expense of collecting its source materials, which essentially is done by hand. Perhaps someone will come along and invent a more efficient method of harvesting agar from its marine source.

Even if you have only the three commonly available components, gelatin, starch, and glycerol, there are many different formulations, each giving a slightly different result.

Equipment

For measuring: You will need a set of measuring spoons and cups. If you are a student and want to make this an "experiment" at your school, you can be more precise by using scales to measure the weights of solid ingredients (like the gelatin) and a graduated cylinder or pipet to measure the liquid ingredients.

The recipes will give you the amounts to use in grams (g) or milliliters (ml), as well as in cups and teaspoons (tsp), so that you can measure your ingredients either way. If you want to experiment with some of your own combinations, you should keep in mind that different ingredients will have different *approximate* conversions:

gelatin, starch, agar, and sorbitol: 3 g = 1 tsp
water: 120 ml = 1/2 cup
glycerol: 3 g = 2.4 ml = 1/2 tsp

For heating: You can use anything from a microwave oven, to a kitchen range, to a hot plate in a school lab. You might also have some way of measuring the temperature of your mixture, which should be heated to just below boiling (95 °C); or you can heat it until it just begins to simmer.

For drying: You can use anything with a nonstick surface, such as a nonstick baking pan, brownie pan, cookie sheet, or muffin tin. You could also use a glass dish if you apply a nonstick spray (there are many household sprays containing teflon that you might use). You can also use a spray on nonstick pans to make removal easier, or if the surface is worn.

Before you begin, it will be useful to make up a "stock" solution of diluted glycerol. Mix up a solution that has 10 ml of glycerol for every liter of water, or 2 tsp of glycerol for every quart. This will give you a solution that is "1% by volume." Every time a recipe calls for a certain amount of a "1% glycerol solution," you can then simply measure from this stock.

Procedure

The procedure is simple, and is basically the same for all of the recipes. You mix the ingredients together, in the amounts given in the recipe, and stir. If you add the solid ingredients to the liquids, your mixture may clump less; however, the order is not that important as long as everything is evenly mixed together.

When additional mixing seems to cause no further dispersion of the components, you then heat the mixture to 95 °C or to the point of initial frothing, whichever comes first. If no thermometer is available, heat until the mixture just begins to simmer. You should stir the mixture during heating. With a microwave oven, you can interrupt the heating periodically, remove the mixture and stir it. If you measure the time it takes a microwave oven to heat a given formulation once, you can then use that time as a good indication of the amount of time required for similar formulations. Heating to too high a temperature will produce frothing, causing the mixture to rise up the sides of the container and boil over.

After heating, stir the hot mixture again well. There should be no visible lumps. Try to avoid forming a lot of froth on the top surface. Excess froth can be scooped out with a spoon and discarded. Very old samples of gelatin or starch may tend to froth more because the polymers slowly degrade over time.

Carefully pour the mixture into a drying pan that has already been placed in a level position. If necessary, you can spread out the mixture to cover the bottom of the pan, either by tilting the pan from side to side, or by smoothing the mixture with a small rod or spoon.

The drying period depends on room temperature and humidity. You can simply leave the pans undisturbed for as long as it takes, which may be several days. You can accelerate the drying by using a mounted hair dryer in a closet-sized room or by putting the pans under a laboratory hood, which simulates a "forced-air oven" environment such as is used in industry. It is easier just to be patient.

You will not be able to remove the sheet easily from the drying pan until it is dry, even when nonstick surfaces are used. The drier the sample, the less it will stick. Formulations that are very much overplasticized will simply produce a gum that cannot be removed in one piece. In some cases, a hard gel will form at first, which can be removed easily, but still contains a significant amount of water. Such a sheet can be easily torn at that stage. Upon drying, the sheet will take on greater strength.

Results may depend on the particular batch of gelatin or starch in the commercial package. Very old samples might be partially degraded, which could affect the quality of the results. Also, if you find that discoloration spots soon develop in a film, the drying pan was not clean, resulting in mold growth. Cleaning the drying pan with soap and water will prevent mold growth, but do not use any abrasive material—the nonstick surface might be damaged.

You should consider this procedure, and the formulations, to be starting points. Feel free to make modifications. Increasing the amount of plasticizer will give a more flexible sheet; less plasticizer increases brittleness. In particularly humid environments, the amount of plasticizer will have to be reduced, by one-half or even more depending on the level of humidity. The procedure and formulations described here represent the simplest way to get reasonably good results with a minimum of effort and time.

Formulations

The first few recipes illustrate the simplest of formulations, containing only gelatin and glycerol. The "A" formulations below are designed for drying in "brownie" pans or dishes, approximately 25 cm x 15 cm. The "B" formulations are for drying in "cookie" trays, approximately 38 cm x 25 cm. The formulations can be scaled up or down to produce larger or smaller sheets, or sheets that are thicker or thinner. Food colors can be added as desired.

Bioglass

This recipe produces a transparent glass-like sheet that can be used in small picture frames as a substitute or replacement for glass or Plexiglas (fig. A.1). It has relatively little plasticizer, so it will not be very flexible; the polymer to plasticizer ratio is about 4:1 by weight. It will hold its shape well, however, and with care you can make it perfectly transparent.

Figure A.1 *Homemade bioglass used as a replacement for glass or Plexiglas in commercial picture frames (5" x 7"; 2 3/8" x 3 3/8"). The bioglass can be made from supermarket gelatin, drugstore glycerol (glycerin), and water. Homemade bioglass illustrates some of the properties of simple bioplastics. (Campus photograph courtesy of Binghamton University)*

A. Combine 12.0 g (4 tsp) gelatin with 240 ml (1 cup) 1% glycerol solution.
B. Combine 36.0 g (12 tsp) gelatin with 480 ml (2 cups) 1% glycerol solution.

For perfect clarity you should use a good nonstick drying surface without a nonstick spray; a nonstick spray will tend to cause small streaks in the sheet. If there are ripples or wrinkles in the final product, there is too much plasticizer. You are aiming for a relatively brittle sheet. You can trim the dried sheet to the required dimensions. Because excessive brittleness can interfere with trimming, you may want to trim the sheet to size before it is thoroughly dry—but still dry enough to remove and handle.

Figure A.1 shows examples of commercial picture frames, one 12.5 cm x 17.8 cm (5" x 7"), and a smaller one 6.0 cm x 8.5 cm (2 3/8" x 3 3/8"), in which the glass covers have been replaced with bioglass.

Laminate

This recipe produces a much more flexible sheet than bioglass, one that can be folded around a small paper or cardboard card (such as a library card, membership card, or identification card) to provide protection from wear and tear. To be flexible the recipe has more plasticizer than the bioglass recipe; the polymer to plasticizer ratio is about 4:3 by weight.

A. Combine 2.25 g (3/4 tsp) gelatin with 135 ml (9/16 cup; i.e., between 1/2 and 5/8 cup) 1% glycerol solution.
B. Combine 6.0 g (2 tsp) gelatin with 360 ml (1 1/2 cup) 1% glycerol solution and 120 ml (1/2 cup) water.

You are aiming for perfect clarity. Trim the dried sheet to slightly larger than twice the size of the card to be laminated. You want to wrap the card as though you were wrapping a gift. Place the card face up on the sheet, lined up along the top edge of the sheet. Fold the sheet from the bottom to cover the card, and trim so that there is about 1/4" excess along the top of the card. Fold the excess over the back side of the card and make a wet seal, as though you were sealing an envelope—wet the surfaces and press together momentarily until a seal is formed. Trim the sides so that there is about 1/4" excess on both sides. On each side carefully cut off the back side of the double layer. Fold each side over and wet-seal.

Biowrap—A Card Protector

This recipe produces a sheet that can be used, for example, to provide a protective envelope for bank cards or credit cards (fig. A.2). It is intermediate in flexibility, not so brittle as bioglass but somewhat less flexible than the biolaminate. The polymer to plasticizer ratio is about 3:2 by

Figure A.2 *Miscellaneous homemade objects: cardholders, buttons, decorations. They can be made from supermarket gelatin, drugstore glycerol (glycerin), and water with optional food-coloring dyes. The items illustrate some of the properties of simple bioplastics.*

weight. Biowrap can also be used for a number of other purposes, including single-use packaging or root-ball wraps.

A. Combine 2.25 g (3/4 tsp) gelatin with 120 ml (1/2 cup) 1% glycerol solution.
B. Combine 6.0 g (2 tsp) gelatin with 320 ml (1 1/3 cup) 1% glycerol solution and 160 ml (2/3 cup) water.

For a card protector, be sure the film is very dry, with no signs of tackiness; otherwise you will have difficulty removing the card from its holder. Place the card horizontally on the sheet, lined up with the top right edge. Fold the sheet from the bottom, over the top of the card, creasing the bottom fold. Trim the sheet, leaving about a 1/4" excess on the top and 1/4" excess on the left. Trim the bottom layer of the left excess, to make it flush with the card. Wet the 1/4" strip on the left, fold it over, and press to make a seal. Make another wet seal with the top 1/4" strip, but do not make the fit too snug. Remove the card. Cut out a small half-moon piece on the open side. Set aside to dry thoroughly. A *very* light dusting of talc (as in baby powder) will help prevent stickiness, but too much may interfere with the functioning of cards with magnetic strips. A bit of technique is required; your second attempt will undoubtedly come out better than your first.

Buttons

This recipe is for a *viscose,* containing a much smaller amount of water than previous recipes. It dries to form a tough, hard solid, that can be drilled, for example, with holes to produce buttons (fig. A.2). The proportions of ingredients required will depend not only on humidity, but also on the particular batch of gelatin in the commercial package. Be prepared to modify the recipe.

Combine 3.0 g (1/2 tsp) glycerol and 12.0 g (4 tsp) gelatin with 60 ml (1/4 cup) hot water.

At this high concentration of gelatin, the mixture will begin to foam and froth well before it reaches 95 °C. Heat only to the initial point of frothing. Wait a few minutes for the froth to collect at the top, then remove it with a spoon. You want to make the buttons with the clear liquid in the bottom of the mixture. You can add food coloring dyes as you wish.

Pour the hot viscose into molds of the desired size and shape. As molds, you can use bottle tops or vial caps, for example. When dry, remove the buttons from the molds and trim as needed. You can use scissors at first, before they are completely hardened. When they are very hard, you can use a small file or sandpaper for touch-up smoothing. Then drill two, or four, holes in the center of each button with a small power tool, or with an awl and round file. For harder buttons use less glycerol.

These buttons will not be highly water resistant, but will survive cold-water handwashing when used, for example, on cardigan-style sweaters. Although they will be somewhat flexible after the washing, they will harden upon drying.

Decorations

You can also use a viscose recipe with molds of other shapes to produce a wide variety of small decorative objects (fig. A.2). You might be able to find small dessert molds available commercially where you live.

This recipe is for somewhat less hard objects than the buttons. Here, too, the best proportions of ingredients may depend on the particular batch of gelatin in the commercial sample you are using; be prepared to make changes.

> Combine 12.0 g (2 tsp) glycerol and 12.0 g (4 tsp) gelatin with 60 ml (1/4 cup) hot water.

You can add food dyes as desired. Remove any foam before pouring into molds, as described for making the buttons. When the material is dry, trim as needed with scissors. You can then use the decorations as wall ornaments, for example, or as tree or plant decorations. You can also place small objects into the molds before pouring the viscose, so as to embed them into the final decoration. Figure A.2 shows some examples.

Varying the Recipes

Some or all of the gelatin can be replaced with other biopolymers, and some or all of the glycerol can be replaced with another plasticizer, such as sorbitol. There are many possible recipes. If, for example, you are business-oriented and cost-conscious, you might wonder how much of the gelatin can be replaced with starch (which is less expensive) and still produce a desirable result.

Other formulations are given below as starting suggestions; each will have slightly different properties.

> ***Gelatin***
>
> A. Combine 0.75 g (1/4 tsp) sorbitol and 2.25 g (3/4 tsp) gelatin with 60 ml (1/4 cup) 1% glycerol solution and 60 ml (1/4 cup) water.
>
> B. Combine 1.5 g (1/2 tsp) sorbitol and 6.0 g (2 tsp) gelatin with 240 ml (1 cup) 1% glycerol solution and 180 ml (3/4 cup) water.

> ***Starch-Gelatin***
>
> A. Combine 1.13 g (3/8 tsp) starch and 1.13 g (3/8 tsp) gelatin with 120 ml (1/2 cup) 1% glycerol solution.
>
> B. Combine 3.0 g (1 tsp) starch and 3.0 g (1 tsp) gelatin with 360 ml (1 1/2 cup) 1% glycerol solution and 120 ml (1/2 cup) water.
>
> *Continued on next page*

Continued from previous page

A. Combine 0.75 g (1/4 tsp) sorbitol, 1.13 g (3/8 tsp) starch and 1.13 g (3/8 tsp) gelatin with 60 ml (1/4 cup) 1% glycerol solution and 60 ml (1/4 cup) water.

B. Combine 1.5 g (1/2 tsp) sorbitol, 3.0 g (1 tsp) starch and 3.0 g (1 tsp) gelatin with 240 ml (1 cup) 1% glycerol solution and 240 ml (1 cup) water.

A viscose: Combine 6.0 g (2 tsp) starch, 6.0 g (2 tsp) gelatin and 3.0 g (2.4 ml) (1/2 tsp) glycerol with 60 ml (1/4 cup) water to produce a viscose.

Starch

Formulations with starch as the only biopolymer form better sheets if a small amount of salt is added. Ammonium acetate works well, but table salt (sodium chloride) can be used as a substitute. If you make up a stock solution of 9 g sodium chloride in a liter of water, 5 ml of the solution will contain 45 milligrams (mg) salt, and 10 ml will contain 90 mg salt. If you have no balance, take a full "pinch" to approximate 90 mg. Or, a solution containing 1 1/2 tsp salt in a quart of water will have approximately 45 mg salt in 1 tsp of solution, and approximately 90 mg salt in 2 tsp of solution.

A. Combine 3.0 g (1 tsp) starch and 45 mg salt with 160 ml (2/3 cup) 1% glycerol solution.

B. Two times A plus 160 ml (2/3 cup) water.

A. Combine 0.75 g (1/4 tsp) sorbitol, 3.0 g (1 tsp) starch and 45 mg salt with 120 ml (1/2 cup) 1% glycerol solution and 40 ml (1/6 cup) water.

B. Two times A plus 160 ml (2/3 cup) water.

Agar

A. Combine 1.5 g (1/2 tsp) agar with 120 ml (1/2 cup) 1% glycerol solution.

B. Two times A plus 180 ml (3/4 cup) water.

A. Combine 0.75 g (1/4 tsp) sorbitol and 1.5 g (1/2 tsp) agar with 60 ml (1/4 cup) 1% glycerol solution and 60 ml (1/4 cup) water.

B. Two times A plus 180 ml (3/4 cup) water.

Starch-Agar

A. Combine 0.75 g (1/4 tsp) starch and 0.75 g (1/4 tsp) agar with 120 ml (1/2 cup) 1% glycerol solution.

B. Combine 3.0 g (1 tsp) starch and 3.0 g (1 tsp) agar with 360 ml (1 1/2 cup) 1% glycerol solution and 60 ml (1/4 cup) water.

A Combine 0.75 g (1/4 tsp) sorbitol, 0.75 g (1/4 tsp) starch and 0.75 g (1/4 tsp) agar with 60 ml (1/4 cup) 1% glycerol solution and 60 ml water.

B. Combine 1.5 g (1/2 tsp) sorbitol, 3.0 g (1 tsp) starch and 3.0 g (1 tsp) agar with 240 ml (1 cup) 1% glycerol solution and 300 ml (1 1/4 cup) water.

Gelatin-Agar

Gelatin-agar mixtures plasticized with glycerol and sorbitol display properties that depend on the relative amounts of the various components.

A. Combine 0.75 g (1/4 tsp) gelatin and 0.75 g (1/4 tsp) agar with 120 ml (1/2 cup) 1% glycerol solution.

B. Three times A plus 60 ml (1/4 cup) water.

A. Combine 0.75 g (1/4 tsp) sorbitol, 0.75 g (1/4 tsp) gelatin and 0.75 g (1/4 tsp) agar with 60 ml (1/4 cup) 1% glycerol solution and 60 ml (1/4 cup) water.

B. Three times A plus 60 ml (1/4 cup) water.

A Combine 0.75 g (1/4 tsp) sorbitol, 1.13 g (3/8 tsp) gelatin and 0.38 g (1/8 tsp) agar with 60 ml (1/4 cup) 1% glycerol solution and 60 ml (1/4 cup) water.

B. Combine 2.25 g (3/4 tsp) sorbitol, 3.0 g (1 tsp) gelatin and 1.5 g (1/2 tsp) agar with 180 ml (3/4 cup) 1% glycerol solution and 240 ml (1 cup) water.

Starch-Gelatin-Agar

Starch-gelatin-agar mixtures can be plasticized with glycerol, sorbitol, or both. Variation in the amounts of all five components allows a range of final properties to be achieved.

Continued on next page

Continued from previous page

A. Combine 0.75 g (1/4 tsp) sorbitol, 1.5 g (1/2 tsp) starch, 0.75 g (1/4 tsp) gelatin and 0.75 g (1/4 tsp) agar with 120 ml (1/2 cup) 1% glycerol solution.

B. Two times A plus 180 ml (3/4 cup) water.

A. Combine 0.75 gram (1/4 tsp) sorbitol, 0.75 g (1/4 tsp) starch, 1.5 g (1/2 tsp) gelatin and 0.75 g (1/4 tsp) agar with 120 ml (1/2 cup) 1% glycerol solution.

B. Two times A plus 180 ml (3/4 cup) water.

A. Combine 1.5 g (1/2 tsp) sorbitol, 1.5 g (1/2 tsp) starch, 1.5 g (1/2 tsp) gelatin and 0.75 g (1/4 tsp) agar with 120 ml (1/2 cup) 1% glycerol solution.

B. Combine 1.5 g (1/2 tsp) sorbitol, 2.25 g (3/4 tsp) starch, 2.25 g (3/4 tsp) gelatin and 1.5 g (1/2 tsp) agar with 240 ml (1 cup) 1% glycerol solution and 180 ml (3/4 cup) water.

"1-2-3 Plastic"

A novel formulation, called here "1-2-3 plastic," is made by combining starch, gelatin, and agar in proportions reflecting their relative cost. Starch, the least expensive, is present in greatest proportion; gelatin in a smaller amount; and agar in the smallest proportion.

A. Combine 0.75 g (1/4 tsp) sorbitol, 1.5 g (1/2 tsp) starch, 0.75 g (1/4 tsp) gelatin and 0.38 g (1/8 tsp) agar with 120 ml (1/2 cup) 1% glycerol solution.

B. Combine 1.5 g (1/2 tsp) sorbitol, 4.5 g (1 1/2 tsp) starch, 2.25 g (3/4 tsp) gelatin and 1.13 g (3/8 tsp) agar with 360 ml (1 1/2 cup) 1% glycerol solution and 60 ml (1/4 cup) water.

The three components represent three different sources of feedstock: a terrestrial plant polysaccharide, a terrestrial animal protein, and a marine plant polysaccharide. The mix has several interesting characteristics.

Because of the widely differing origins of the components, discarded material would provide a particularly diverse mix of substrates for microorganisms in a variety of waste environments. Depending on where the discarded material ends up, one or the other of the components will provide the best substrate for the initial steps of degradation.

Figure A.3 *A homemade root-ball wrap made of "1-2-3 plastic," containing starch, gelatin, agar, glycerol (glycerin), sorbitol, and water. The root-ball wrap can be placed in the ground together with the plant. The root-ball wrap biodegrades. It can also be made with the simpler "biowrap" formulation.*

Moreover, the relative amounts of carbohydrate and protein approximate compositions that satisfy the nutritional requirements for farm animals, as described in Chapter 7. Fast-food trays, together with scrap food, could be pasteurized and used to produce farm animal feed.

The same diversity of composition might prove to be a particularly useful blend of nutrients for soil enrichment in agricultural or garden applications. Figure A.3 shows a root-ball wrap made of 1-2-3 plastic. Subsequently, the plant shown in figure A.3 was planted in upstate New York. Two months later the plant was examined and the wrap had disappeared without a trace. The plant was doing fine.

Other Possibilities

Other biopolymers are available through chemical supply houses. Some, such as carrageenan, are soluble in hot water. Cellulose, chitin, and casein are not directly soluble in hot water, but they form slurries with gelatin or starch, which can be dried to form sheets. With some fancier chemistry they can be made to be soluble, which is perhaps best left to trained chemists. As mentioned in Chapter 7, starch-cellulose and starch-casein composites have been suggested as bioplastic formulations for packaging applications.

In addition to glycerol and sorbitol, glucose and sucrose have been used as plasticizers. There may be other common compounds that have not even been discovered to be good plasticizers.

Other components can be used to impart desirable properties to the bioplastics. These might be simple cosmetic additives, such as coloring dyes, or they might serve an important purpose, such as a salt to improve sheet quality, or a fertilizer in a garden or agricultural application. A component might be mixed throughout the bioplastic composition, or it might be added externally as a coating. Coatings that provide a water barrier are particularly important. Fillers can be sought to make formulations less expensive. The formulations described above have illustrated just a few of the possibilities.

Standard Tests

The physical properties of plastics determine their useful applications. It is important to have standard procedures for determining those properties, and in the United States it is the American Society for Testing and Materials (ASTM) that develops detailed measurement procedures (Chapter 5). Adherence to those procedures allows uniformity within the plastics industry, and provides a basis for comparison of one product with another.

The growing importance of biodegradability has required new procedures for evaluating that property as well, and ASTM has been establishing suitable tests. Because biological processes are involved, establishing biodegradability test procedures is more complex than establishing procedures for measuring physical properties.

Some of the standard physical properties that are often used to characterize plastics are glass temperature, tensile strength, indentation hardness, elongation, loss of volatile components, barrier to oxygen, barrier to moisture, oil resistance, ease of marring, fold resilience, and tear resistance.

You can measure some physical properties easily, and you can practice on some of the homemade bioplastic materials described here. Examples of such properties are clarity, flexibility at refrigerator temperatures (~5 °C), flexibility at freezer temperatures (~ –20 °C), and the effect of exposure to 800-watt microwave radiation, typical of home microwave ovens. Also of interest is how those properties change with composition. Simple experiments aimed at demonstrating the different behavior of those materials are easy to carry out.

Precise quantitative measurements of other properties may require equipment that is complex and expensive. Instruments for measuring glass temperature and tensile strength accurately, for example, can cost thousands of dollars. On the other hand, approximate measures of some properties can be obtained with simpler equipment. Two interesting and important properties are described here in greater detail.

Glass Temperature

The **glass transition** of a polymer material refers to the heat-induced change from a brittle, glassy solid to a flexible, pliable solid, or the reverse change induced by cooling. The **glass transition temperature**, or **glass temperature, T_g**, is the temperature at which that change occurs. The T_g of a polymer material can be measured accurately with specially designed instruments. If a refrigerator and freezer are available, however, you can at least determine whether the T_g of a sample is greater or less than the refrigerator temperature, and greater or less than the freezer temperature. For example, if the sample is flexible at a room temperature of, say, 20 °C, but becomes brittle at a refrigerator temperature of 5 °C, the T_g of the sample lies between those two temperatures. If the sample is flexible at 5 °C, but becomes brittle at a freezer temperature of –20 °C, the T_g lies between those two temperatures. If the sample remains flexible even at a freezer temperature of –20 °C, the T_g is below that temperature. Glycerol-plasticized bioplastics tend to have glass temperatures lower than sorbitol-plasticized bioplastics.

Common polymers are made flexible below their glass temperatures with the use of plasticizers. Ordinary paper, for example, is plasticized by water. If paper is heated in a drying oven near 110 °C, it becomes brittle because cellulose has a glass temperature of around 225 °C.

One scientific instrument that allows precise measurement of glass temperatures is a differential scanning calorimeter (DSC). It is expen-

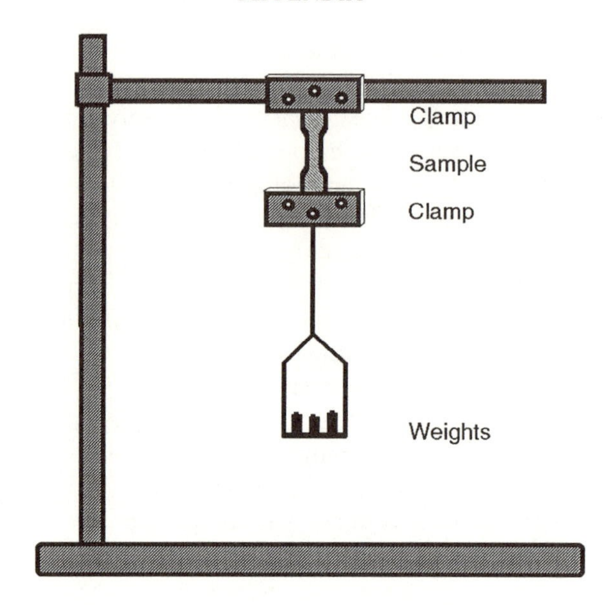

Figure A.4 *A simple tensile-strength measuring device*

sive, costing thousands of dollars, but colleges and universities often have such instruments for scientific research, and sometimes for student instruction.

Tensile Strength

Accurate measurements of tensile strength require expensive equipment and careful adherence to standard protocols, such as ASTM Standard D-1708. Qualitative observations can indicate only roughly whether a sample has a very low tensile strength or a significantly large tensile strength.

Figure A.4 shows a simple device for estimating tensile strength that can be constructed relatively easily. It can at least give some idea of the relative strengths of two samples, if the same procedure is used for both tests.

The device can be constructed, for example, from a laboratory ringstand to which the top horizontal supporting rod is attached. The support for the weights might be a weight pan from a commercial analytical balance. For an even simpler device the weight pan and weights could be replaced with a spring scale.

Figure A.5 *A template for preparing tensile-strength samples*

The two sample clamps can be machined easily from four flat pieces of stock, drilled with holes for fastening one pair to the top support rod, and for supporting the weight pan from the bottom pair. Two additional holes in each pair, together with four small bolts and nuts, allow the clamps to be tightened after a sample is inserted.

A sample is clamped in place, and weights are added slowly to the pan. The weights provide an increasing stress. The sample will become taut, and may elongate to some degree. It will eventually break. The result of this simple procedure would more correctly be called the tensile strength at break, or the fracture strength. The point at which the sample breaks depends on the rate at which the stress is applied, and a complete characterization of tensile strength takes that factor into account.

The added weights will provide a rough indication of the tensile strength. If the weights are added slowly, with short time intervals between additions, and the same procedure is used for each successive test, the results will have some significance. How reproducible the procedure is can be tested by repeating the measurement with two, or more, identically prepared samples.

Probably the most important part of the measurement is preparing the sample and securing it in the holding clamps. Figure A.5 provides a dog-bone-shaped template with which to prepare samples. Use a copy of the template and a razor blade or scalpel to cut the sample into the desired shape. It is very important for each sample to have the same shape. The most important dimension in the shape is the width in the narrowest region of the sample, which is 0.25 inch (0.64 cm) in the design shown in figure A.5.

The measurements will not be exactly reproducible. Even measurements made with elaborate equipment are often not as reproducible as one might hope for. The simple device shown in figure A.4 will be able to distinguish differences in the tensile strengths of two samples, especially if those differences are large. The tensile strength of some bioplastics can be affected by the atmospheric humidity, because of their sensitivity to water vapor. The tensile strength of a sample might vary from day to day for this reason. To eliminate such variation, accurate measurements are made under conditions of controlled humidity. That control, however, is not easy to achieve.

It is also necessary to measure the thickness of the sample as accurately as possible. A micrometer is a device for carrying out such measurements. Many micrometers have their scales indicated in inches; the scale can be converted to centimeters by multiplying the value in inches by the factor 2.54. Also, the unit of "mils" is often used, where one mil equals 0.001 inch.

1.00 mil = 0.00100 inch = 0.00254 cm
2.00 mil = 0.00200 inch = 0.00508 cm

The tensile strength is calculated by first calculating the cross-sectional area of the segment of the sheet that broke. The cross-sectional area is calculated as the width (0.25 inch, or 0.64 cm) multiplied by the thickness.

For example, if the sample has a thickness of 0.0030 inch (0.00762 cm), the cross-sectional area, in units of cm^2, is:

$$0.64 \text{ cm} \times 0.00762 \text{ cm} = 0.0049 \text{ cm}^2$$

The tensile strength is then calculated by dividing the weight added at the break point by that cross-sectional area.

For example, if the sample described above breaks after a final weight is added, which brings the total weight to 500 grams, or 0.500 kilograms (kg), the tensile strength (t.s.) is calculated as:

$$\text{t.s.} = 0.500 \text{ kg} / 0.0049 \text{ cm}^2 = 102 \text{ kg/cm}^2$$

Other units for tensile strength are pounds per square inch (psi) and megapascals (MPa). The conversion factors for these units are:

1 psi = 0.0703 kg/cm^2	1 kg/cm^2 = 14.22 psi
1 MPa = 10.20 kg/cm^2	1 kg/cm^2 = 0.0981 MPa
1 psi = 0.00689 MPa	1 MPa = 145.0 psi

In the illustration example, the 102 kg/cm^2 tensile strength is equivalent to 1450 psi or 10.0 MPa.

Designing Science Projects

You may want to design your own project—a simple or an elaborate one. For example, you may want to develop sheets that have large tensile strengths. Very thin sheets, containing relatively little plasticizer, will have larger tensile strengths, other factors being equal. Keep in mind that tensile strength may change in time. You may want to develop a sheet that is microwave resistant, or look for sheets that remain flexible at as low a temperature as possible.

You may be more interested in measuring how degradable bioplastics are. Test environments can be devised from simple materials. The simplest would be nothing more than a container of water. Moist soil can be used, or compost, or marine water, or simulations of landfill environments. Control tests can be carried out with other materials at the same time to provide a comparison, including materials that may be less degradable, as well as materials that you expect will be equally degradable, like food. Figure A.6 shows the results of a degradation experiment using gelatin-glycerol button-shaped test samples. The samples shown were originally buried outdoors in garden soil approximately six inches deep (with the burial location well marked!). The elapsed time was twelve days, but the degradation rate in any particular experiment will depend on many factors (Chapter 5).

More elaborate projects might include varying the amount of moisture or the amount of exposure to sunlight—or determining how periodic mixing affects the rate of degradation. Some tests are listed in Chapter 5, including the **soil test**, **Sturm test**, **earthworm test,** and **cress seed test**; another is a simple **mold test**.

You may want to experiment simply to satisfy your own curiosity, or as a hobby. Students may want to carry out science projects for class work, a science fair, or even a competition. In any case, you will get some idea of what it is like to do scientific research on a topic of broad

Figure A.6 *A biodegradation experiment using homemade buttons.*

practical interest. And you will have the chance to examine these intriguing new materials for yourself. If you become serious about doing experiments, be sure to keep a notebook. It is nearly impossible to recall formulations from memory, and—who knows—maybe you will come across something *really* interesting.

Happy experimenting!

Notes

(*op. cit.* refers to the **Reading List**, which follows the **Glossary**)

PART ONE PLASTICS

Chapter 1 The Age of Plastics

(pp. 3–4) For accounts of the contribution of plastics to shaping twentieth-century culture, read *American Plastic: A Cultural History*, by Jeffrey L. Meikle (Rutgers University Press, New Brunswick, New Jersey, 1995), and *Plastic: The Making of a Synthetic Century*, by Stephen Fenichell (HarperCollins, New York, 1996).

The life-supporting role of plastic in water-deprived areas of Africa is described by Ryszard Kapuscinski, as recounted by Stephen Fenichell, in *Plastic: The Making of a Synthetic Century,* ibid., pp. 9–10.

(pp. 5–6) Writers on plastics have invariably cited two events. One is the instantaneous popularity of the "Plastics!" scene in the movie *The Graduate*, which has become an American cultural icon. The second is introduction of the term *Age of Plastics.* It was coined in the 1920s by plastics industry promoters, but evolved into a more meaningful label once global plastics production exceeded steel production in volume, around 1979. Plastic is approximately one-seventh as dense as steel, and the weight of plastics produced now exceeds one-seventh the weight of steel produced.

(p. 6) The 1986–1998 data for figure 1.1 are from the Society of the Plastics Industry (SPI) as cited in *Chemical & Engineering News*, 23 June 1997, p. 43; and 28 June 1999, p. 37. Those data do not include some engineering resins and others; 1998 production of those totaled 11.3 billion pounds (SPI). 1999 and 2000 data are from the American Plastics Council (APC) as cited in *Chemical & Engineering News*, 26 June 2000, p. 55; 25 June 2001, p. 47, and 3 September 2001, p. 6; for consistency, estimates for thermoplastic polyesters and non-epoxy thermosetting resins, taken from 1998 and 1999, were added.

(pp. 6–7) The data for figure 1.2 are from the Society of the Plastics Industry, as cited in *Chemical & Engineering News*, 5 October 1998, p. 17.

Packaging is sometimes categorized as follows: packaging at the point of sale (primary; 57 percent), group packaging (secondary; 17 percent), and transport packaging (tertiary; 26 percent). See F. Mader, in *Chemical As-*

pects of Plastics Recycling (1997), *op. cit.*, pp. 13–27. Plastics packaging statistics are from the same source, and also from *Chemical & Engineering News*, 5 October 1998, p. 17.

Other categorizations are packaging in contact with the product (primary), packaging for a product that contains one or more primary packages (secondary), and packaging that contains one or more secondary packages (tertiary), from *Industrial Ecology*, by T. E. Graedel and B. R. Allenby (Prentice-Hall, Englewood Cliffs, New Jersey, 1995), p. 400.

For more on food packaging trends, see *Food Packaging, Testing Methods and Applications* (2000), *op. cit.*

(p. 8) Trends in packaging adhesives are described in *Chemical & Engineering News*, 20 April 1998, pp. 30–32.

Chapter 2 Plastics as Materials

(p. 10) For additional information on materials and materials science, see *Materials Chemistry: An Emerging Discipline*, ed. L. V. Interrante, L. A. Caspar, and A. B. Ellis (American Chemical Society, Washington, D.C., 1995); *Stuff: The Materials the World Is Made Of*, by I. Amato (BasicBooks, New York, 1997); and *Made to Measure: New Materials for the 21st Century*, by P. Ball (Princeton University Press, Princeton, 1997).

(p. 12) Early, and modern, fiber-reinforced composites are described by B. Parkyn, in *The Development of Plastics*, ed. S.T.I. Mossman and P.J.T. Morris (Royal Society of Chemistry, Cambridge, 1994), pp. 105–114.

Chapter 3 Plastics and the Environment

(p. 15) The plastics feedstock figure of 4 percent is from R. S. Stein, in *Plastics, Rubber, and Paper Recycling: A Pragmatic Approach* (1995), *op. cit.*, p. 27. A figure of 5 percent and the processing energy requirement of 2 percent are cited by J. Baldwin, in *Chemical Aspects of Plastics Recycling* (1997), *op. cit.*, pp. 95–105.

(pp. 15–16) Fossil fuel supply estimates vary. In its 2000 assessment the United States Geological Survey estimated a world oil reserve of 215 billion tons; see *Science* 289 (14 July 2000): 237. See also *Environmental Science: The Natural Environment and Human Impact*, by A. R. Jackson and J. M. Jackson (Longman, Essex, England, 1996), pp. 247–249; *Environmental Science: An Introduction* (2d edition), by G. Tyler Miller Jr. (Wadsworth Publishing Company, Belmont, California, 1988), pp. 232–267; *Beyond the Limits*, by D. H. Meadows and D. L. Meadows (Earthscan Publications, London, 1992); and *Our Changing Planet: An Introduction to Earth System Science and Global Environmental Change* (2d edition), by F. T. Mackenzie (Prentice-Hall, Englewood Cliffs, New Jersey, 1998), pp. 231–234.

(p. 16) See the **Glossary** for terms related to **waste**, including **garbage**, **refuse**, **rubbish**, **solid waste**, **trash**, **waste stream**, and **yard waste**.

The 1970 estimate of the plastics waste stream comes from W. A. Mack, in *Disposal of Plastics with Minimum Environmental Impact*, ASTM Special Technical Publication 533 (ASTM, West Conshohocken, Pennsylvania, 1973).

(pp. 16–17) United States MSW statistics are from the U.S. Environmental Protection Agency Fact Sheet, *Municipal Solid Waste Generation, Recycling and Disposal in the United States: Facts and Figures for 1998* (U.S.E.P.A., April 2000).

Statistical data on plastics waste are taken from *Rubbish! The Archaeology of Garbage*, by W. Rathje and C. Murphy (HarperCollins, New York, 1992), pp. 47, 99–102; W. Pearson, in *The McGraw-Hill Recycling Handbook*, ed. H. F. Lund (McGraw-Hill, New York, 1993), pp. 14.1–14.32; R. Narayan, in *Polymers from Agricultural Coproducts* (1994), *op. cit.*, pp. 2–28; C. P. Rader and R. F. Stockel, in *Plastics, Rubber, and Paper Recycling: A Pragmatic Approach* (1995), *op. cit.*, pp. 2–10; F. Mader, in *Chemical Aspects of Plastics Recycling* (1997), *op. cit.*, pp. 13–27, 199–201, 214; "Evaluation of U.S. Grains Council Bioplastic Programs and the Markets in Japan, Korea, and Taiwan," Preliminary Report, U.S. Grains Council, 1999; *BioCycle* 40, 8 (August 1999): 8; and the U.S. Environmental Protection Agency Fact Sheet, *Municipal Solid Waste Generation, Recycling and Disposal in the United States: Facts and Figures for 1998* (U.S.E.P.A., April 2000).

(p. 17) The estimate of plastics waste generated at sea is from R. Narayan, in *Polymers from Agricultural Coproducts* (1994), *op. cit.*, p. 6.

Litter is characterized, anecdotally, in *Polymers and Ecological Problems*, ed. J. Guillet (Plenum, New York, 1973). The Texas cleanup episode is recounted in *The Chemistry of Polymers* (2d edition), by J. W. Nicholson (Royal Society of Chemistry, Cambridge, 1997), p. 176.

(p. 18) The New York City waste-management statistic is cited in *BioCycle* 39, 7 (July 1998): 74; and 42, 4 (April 2001): 22–23.

The reference to plastics restrictions in Tokyo is from *Plastics Waste Recovery of Economic Value*, by J. Leidner (Marcel Dekker, New York, 1981), p. iii.

(p. 19) The relative energy requirements for various materials production are described by J. Guillet, in *Degradable Polymers: Principles and Applications* (1995), *op. cit.*, pp. 224–226.

J. Bickerstraffe presents the complexities of analyzing the environmental impact of packaging in *Chemical Aspects of Plastics Recycling* (1997), *op. cit.*, pp. 3–12. See also "Waste Stream Case Study 1: Packaging," by B. Goeke and F. Chalot, in Washington Waste Minimization Workshop, Vol. I (Organisation for Economic Co-operation and Development, Paris, 1996), pp. 11–102.

Martin B. Hockings's comparison of a hot-drink paper cup with a polystyrene plastic-foam cup is found in *Science* 251 (1 February 1991): 504–505; *Nature* 369 (12 May 1994): 107; ibid., 371 (6 October 1994): 481–482. Also see *Industrial Ecology*, by T. E. Graedel and B. R. Allenby (Prentice-Hall, Englewood Cliffs, New Jersey, 1995), pp. 149–150.

The packaging reduction statistics come from the American Plastics Council.

(p. 20) The preconsumer plastics recycling statistic is from W. Pearson, in *The McGraw-Hill Recycling Handbook*, ed. H. F. Lund (McGraw-Hill, New York, 1993), pp. 14.1–14.32.

The 3M program is mentioned in Fisher Scientific's *Lab Reporter*, July/August 1997.

The effects of recycling on polystyrene properties are described by H. Djidjelli and D. Benachour, in *"Petro" Polymers vs. "Green" Polymers* (1998), *op. cit.,* pp. 181–185.

(p. 21) The collection statistics are from *BioCycle* 39, 2 (February 1998): 96. The recycling program statistics are from "The State of Garbage in America," by J. Glenn, *BioCycle* 40, 4 (April 1999): 60–71.

The poly(ethylene terephthalate) statistic is from the American Plastics Council, as reported in *Chemical & Engineering News*, 10 November 1997, p. 9.

Western European recycling data are given by F. Mader, in *Chemical Aspects of Plastics Recycling* (1997), *op. cit.*, pp. 13–27.

(p. 22) The references to ˙ncineration in Western Europe are from J. M. O'Neill, in *Chemical Aspects ɔf Plastics Recycling* (1997), *op. cit.*, pp. 28–40; and from M. de Bertoldi, *BioCycle* 39, 6 (June 1998): 74–75.

(pp. 22–23) A cracking-plant study by BP Chemical is reported in *Chemical & Engineering News*, 28 September 1998, p. 13. For more on polymer cracking, see a report by J. H. Brophy, S. Hardman, and D. C. Wilson, in *Chemical Aspects of Plastics Recycling* (1997), *op. cit.*, pp. 214–219.

(pp. 23–24) For more on methane production in landfills, and landfills in general, see *Rubbish! The Archaeology of Garbage*, by W. Rathje and C. Murphy (HarperCollins, New York, 1992), pp. 108–112.

New York State's landfill count is from the NYS Legislative Commission of Solid Waste Management 1998 Report, "Where Will All the Garbage Go?" cited in *BioCycle* 39, 12 (December 1998): 10.

New York City's waste-disposal cost estimate is from the city's Comprehensive Solid Waste Management Plan Draft Modifications, 3 April 1998, cited by D. Biddle, in *BioCycle* 39, 7 (July 1998): 72–79. See also *BioCycle* 42, 4 (April 2001): 22–23.

(p. 25) The estimates of biomass production are taken from R. Narayan, in *Polymers from Agricultural Coproducts* (1994), *op. cit.,* p. 2.

Biomass to energy conversion is described in "Fuels and Chemicals from Biomass," ed. B. C. Saha and J. Woodward, American Chemical Society Symposium Series 666 (American Chemical Society, Washington, D.C., 1997). See also "Roles for Biomass Energy in Sustainable Development," by R. Williams, in *Industrial Ecology and Global Change*, ed. R. Socolow, C.

Andrews, F. Berkhout, and V. Thomas (Cambridge University Press, Cambridge, 1994), pp. 199–225.

(p. 27) The relative costs of accepting composting feedstocks in noncompostable bags versus compostable bags is described by R. Tyler in *BioCycle* 39, 3 (March 1998): 58–61. See also *BioCycle* 40, 5 (May 1999): 55–56.

The effects of plastics in compost are described, anecdotally, by J. Purman, in *BioCycle* 39, 3 (March 1998): 62–63.

An anaerobic composting project is described in *BioCycle* 41, 9 (September 2000): 35.

(p. 28) The references to composting in Western Europe are from M. de Bertoldi, *BioCycle* 39, 6 (June 1998): 74–75.

The organic waste statistics come from the U.S. Environmental Protection Agency Fact Sheet, *Municipal Solid Waste Generation, Recycling and Disposal in the United States: Facts and Figures for 1998* (U.S.E.P.A., April 2000).

(pp. 28–29) Reports on food residuals composting can be found in *BioCycle* 39, 8 (August 1998): 50–60, by N. Goldstein, J. Glenn, and K. Gray; and in *BioCycle* 41, 8 (August 2000): 48–54, by N. Goldstein, D. Block, and C. Ochins.

(p. 29) The 1970 waste estimates are from G. L. Huffman and D. J. Keller, in *Polymers and Ecological Problems, Polymer Science and Technology*, Vol. 3, pp. 155–167, ed. J. Guillet (Plenum, New York, 1973).

(pp. 29–30) The data for figures 3.1 and 3.2 are from the U.S. Environmental Protection Agency Fact Sheet, *Municipal Solid Waste Generation, Recycling and Disposal in the United States: Facts and Figures for 1998* (U.S.E.P.A., April 2000).

Chapter 4 The Chemical Nature of Plastics

There is a very large literature on polymers and plastics. *Plastics Materials* (7th edition), by J. A. Brydson (Butterworth-Heinemann, Oxford, 1999), remains a milestone in reference works on the subject. Introductions are *An Introduction to Polymer Science*, by H.-G. Elias (Wiley-VCH, Weinheim, Germany, 1997); *The Chemistry of Polymers* (2d edition), by J. W. Nicholson (Royal Society of Chemistry, Cambridge, 1997); *Polymer Chemistry: An Introduction* (3d edition), by M. P. Stevens (Oxford University Press, Oxford, 1998); and *The Elements of Polymer Science and Engineering* (2d edition), by A. Rudin (Academic Press, San Diego, California, 1999).

The early history of polymer science is described in *Inventing Polymer Science: Staudinger, Carothers, and the Emergence of Macromolecular Chemistry*, by Yasu Furukawa (University of Pennsylvania Press, Philadelphia, 1998).

(p. 31) The term **polymer** was originally used in chemistry to refer to compounds consisting of the same elements in the same weight proportion, but having different molecular weights and different properties; e.g., acetylene,

C_2H_2, and benzene, C_6H_6, were said to be *polymeric* with each other. The chemical nature of what are now called polymers was not discovered until the 1920s.

Organic polymers are emphasized here. There are also *inorganic* polymers, such as siloxanes that contain the element silicon (Si).

(pp. 39–40) A thorough account of additives is *Plastics Additives: An A–Z Reference*, ed. Geoffrey Pritchard (Chapman & Hall, New York, 1998). A brief report on the plastics additives industry appeared in *Chemical & Engineering News*, 4 December 2000, pp. 21–31.

(p. 44) More information on PET production developments is found in *Chemical & Engineering News*, 27 July 1998, pp. 33–35, and 22 May 2000, pp. 25–26.

Thermoplastics production statistics are from the Society of the Plastics Industry, cited in *Chemical & Engineering News*, 28 June 1999, p. 37.

(p. 45) The United States SPI Resin Identification Code, the use of which does not imply a guarantee of recycling, thereby differs from the "Green Dot" in Germany (Chapter 9), which does.

The International Standards Organization (ISO) has developed a set of recommendations for marking plastic parts that includes much more detailed information than the SPI code.

New engineering plastics are described in *Chemical & Engineering News*, 5 October 1998, pp. 18–21.

(p. 46) The biodegradation of poly(vinyl alcohol), for example, is described by S. Matsumura and K. Toshima, in *Hydrogels and Biodegradable Polymers for Bioapplications* (1996), *op. cit.*, pp. 137–148.

(p. 47) For more on biomedical applications, see *Biomaterials Science: An Introduction to Materials in Medicine*, ed. B. D. Ratner, A. S. Hoffman, F. J. Schoen, and J. E. Lemons (Academic Press, San Diego, California, 1996).

(p. 50) Fiber production statistics are from *Chemical & Engineering News*, 15 May 2000, p. 25.

Rubber processing is described by J.W.M. Noordermeer, in *"Petro" Polymers vs. "Green" Polymers* (1998), *op. cit.*, pp. 131–139.

Chapter 5 Plastics Degradation

(p. 52) Three categories of plastics degradability are also used in "Degradable Polymers," by A.-C. Albertsson, *Journal of Macromolecular Science, Pure and Applied Chemistry* A30 (1993): 757–765. Controlled degradation is discussed by A.-C. Albertsson and S. Karlsson, in *Macromolecular Design of Polymeric Materials*, ed. K. Hatada, T. Kitayama, and O. Vogl (Marcel Dekker, New York, 1997), pp. 793–802. For a brief overview, see also "The Design of Polymers with Controlled Lifetimes," by G. Scott, ibid., pp. 803–811.

(p. 55) Early reports on the photoactivated degradation of plastics are given by J. E. Guillet, in *Polymers and Ecological Problems, Polymer Science and*

Technology, Vol. 3, ed. J. Guillet (Plenum, New York, 1973), pp. 1–25; by G. Scott, ibid., pp. 27–44; and by B. Baum and R. A. White, ibid., pp. 45–60.

Ethylene-carbon monoxide copolymers are described by G. Harlan and C. Kmiec, in *Degradable Polymers: Principles and Applications* (1995), *op. cit.,* pp. 153–168, and by J. Guillet, ibid., p. 230.

(p. 57) For one long-term (six-year) study of degradable agricultural covers, see S. R. Yang and C. H. Wu, in *Macromolecular Symposia* 144 (1999): 101–112.

(pp. 60–63) For a more detailed discussion of chemical kinetics, see, for example, *Physical Chemistry*, 2d edition, by R. S. Berry, S. A. Rice, and J. Ross (Oxford University Press, Oxford, 2000).

(pp. 64–67) More detailed descriptions of biodegradation are given by D. L. Kaplan, J. M. Mayer, D. Ball, J. McCassie, A. L. Allen, and P. Stenhouse, in *Biodegradable Polymers and Packaging* (1993), *op. cit.*, pp. 4–10; and by R. P. Wool, in *Degradable Polymers: Principles and Applications* (1995), *op. cit.*, pp. 138–152.

(p. 66) A report on the glacial iceman, with photograph, is found in *Science* 289 (29 September 2000): 2253–2254.

The presence of old undecomposed food items and newspapers in landfills, and many other examples of old, but undegraded, organic matter, are described in *Rubbish! The Archaeology of Garbage*, by W. Rathje and C. Murphy (HarperCollins, New York, 1992).

(p. 67) Human impacts on the carbon cycle are evaluated by R. U. Ayres, W. H. Schlesinger, and R. H. Socolow, in *Industrial Ecology and Global Change*, ed. R. Socolow, C. Andrews, F. Berkhout, and V. Thomas (Cambridge University Press, Cambridge, 1994), pp. 121–155. See also a report by B. Hileman, in *Chemical & Engineering News*, 9 August 1999, pp. 16–23.

(p. 73) The cooperative efforts to establish definitions and protocols related to plastics degradation are described by K. J. Seal, in *Chemistry and Technology of Biodegradable Plastics* (1994), *op. cit.*, pp. 116–134; G. Swift, in *Ecological Assessment of Polymers: Strategies for Product Stewardship and Regulatory Programs* (1997), *op. cit.*, pp. 291–306; and D. Riggle, in *Moving Towards Consensus on Degradable Plastics* (1998), *op. cit.*

The workings of the ASTM, CEN, and BPS are summarized, respectively, by R. Narayan, in *Biodegradable Polymers and Plastics* (1992), *op. cit.*, pp. 176–187; P. Breant and Y. Aitken, ibid., pp. 165–168; and K. Fukuda, ibid., pp. 169–175.

(p. 75) Some of the complexities of measuring plastics degradation are described by R. Bartha, A. V. Yabannavar, M. A. Cole, and J. D. Hamilton, in *Ecological Assessment of Polymers: Strategies for Product Stewardship and Regulatory Programs* (1997), *op. cit.*, pp. 167–184. See also the study by J. M. Mayer, D. L. Kaplan, R. E. Stote, K. L. Dixon, A. E. Shupe, A. L. Allen, and J. E. McCassie, in *Hydrogels and Biodegradable Polymers in Bioapplications* (1996), *op. cit.*, pp. 159–170.

The summary of a Tier 3 report on eleven polymer materials is given by G. Croteau, in *BioCycle* 39, 3 (March 1998): 71–75.

For an example of developing laboratory-scale composting test methods to determine polymer biodegradability, applied to cellulose acetate, see work by R. A. Gross, J.-D. Gu, D. Eberiel, and S. P. McCarthy, in *Journal of Macromolecular Sciences, Pure and Applied Chemistry* A32 (1995): 613–628.

(p. 77) The compostable logo was reported in *BioCycle* 41, 8 (August 2000): 46.

PART TWO BIOPLASTICS

Chapter 6 Biopolymers

(pp. 84–85) Overviews on cellulose are given by R. M. Rowell, in *Emerging Technologies for Materials and Chemicals from Biomass* (1992), *op. cit.*, pp. 12–27; D. N.-S. Hon, ibid., pp. 176–196; and R. D. Gilbert and J. F. Kadla, in *Biopolymers from Renewable Resources* (1998), *op. cit.*, pp. 47–95.

(p. 85) Microbially produced cellulose is described by D. Byrom, in *Biomaterials: Novel Materials from Biological Sources* (1991), *op. cit.*, pp. 263–283; and by D. L. Kaplan, in *Biopolymers from Renewable Resources* (1998), *op. cit.*, pp. 89–90.

(pp. 85–86) A summary description of the properties and applications of starch is given by R. L. Shogren, in *Biopolymers from Renewable Resources* (1998), *op. cit.*, pp. 30–46.

The complexities of starch structure-function relationships are described, for example, in *Starch: Structure and Functionality*, ed. P. J. Frazier, P. Richmond, and A. M. Donald (Royal Society of Chemistry, Cambridge, 1997). Also see reports by R. B. Friedman, D. J. Mauro, R. J. Hauber, and F. R. Katz, in *Carbohydrates and Carbohydrate Polymers* (1993), *op. cit.*, pp. 62–72; and by J.-L. Jane, J. J. Shen, M. Radosavljevic, T. Kasemsuwan, A. Xu, and P. A. Seib, ibid., pp. 174–184.

(p. 88) See a summary description of chitin and chitosan by S. M. Hudson and C. Smith, in *Biopolymers from Renewable Resources* (1998), *op. cit.*, pp. 96–118. See also *Chitin and Chitosan—An Expanding Range of Markets Awaits Exploitation*, Reports Group, Technical Insights, 3d edition (Wiley, New York, 1998).

Recent efforts to extract chitin from shellfish waste in Maryland and market it are described by D. Block, in *BioCycle* 41, 12 (December 2000): 30–33.

(pp. 88–89) See reports on agar, by H. H. Selby and R. L. Whistler, in *Industrial Gums: Polysaccharides and Their Derivatives*, 3d edition, ed. R. L. Whistler and J. N. BeMiller (Academic Press, New York, 1993), pp. 87–103; alginates, by K. Clare, ibid., pp. 105–143; and carrageenan, by G. H. Therkelsen, ibid., pp. 145–180. For more on alginates, see a report by D. E. Day, in *Biopolymers from Renewable Resources* (1998), *op. cit.*, pp. 119–143. For more

on carrageenan, see a report by S. Ramakrishnan and R. K. Proud'homme, in *Polymers from Renewable Resources: Polysaccharides and Agroproteins* (2001), *op. cit.,* pp. 86–101.

For more on the use of polysaccharides in forming microspheres, see C. Thies, in "Polymers in Medicine and Pharmacy," Materials Research Symposium Proceedings, Vol. 394, ed. A. G. Mikos, K. W. Leong, M. J. Yaszemski, J. A. Tamada, and M. L. Radomsky (Materials Research Society, Pittsburgh, Pennsylvania, 1995), pp. 49–54.

(p. 90) Xanthan and other microbial polysaccharides are described by J. D. Linton, S. G. Ash, and L. Huybrechts, in *Biomaterials: Novel Materials from Biological Sources* (1991), *op. cit.*, pp. 215–261; and by W. F. Fett, S. F. Osman, M. L. Fishman, and K. Ayyad, in *Agricultural Materials as Renewable Resources: Nonfood and Industrial Applications* (1996), *op. cit.*, pp. 76–87.

(p 90–91) Lignin is reviewed by R. A. Northey, in *Emerging Technologies for Materials and Chemicals from Biomass* (1992), *op. cit.,* pp. 146–175; and by D. S. Argyropoulos and S. B. Menachem, in *Biopolymers from Renewable Resources* (1998), *op. cit.*, pp. 292–322. See also the comprehensive work "Lignin: Historical, Biological, and Materials Perspectives," American Chemical Society Symposium Series 742, ed. W. G. Glasser, R. A. Northey, and T. P. Schultz (American Chemical Society, Washington, D.C., 2000); and *Biobased Industrial Products* (2000), *op. cit.,* pp. 81–85.

(p. 93) The casein fragment sequence is from R. Osterberger, *Acta Chemica Scandinavica* 18 (1964): 795–804, as cited in *Atlas of Protein Sequence and Structure* (1969), Vol. 4, ed. M. O. Dayhoff, p. D189.

Soy protein is described by Y.T.-P. Ly, L. A. Johnson, and J. Jane, in *Biopolymers from Renewable Resources* (1998), *op. cit.*, pp. 144–176.

(pp. 94–95) Polyhydroxyalkanoates are described by A. Steinbüchel in *Biomaterials: Novel Materials from Biological Sources* (1991), *op. cit.*, pp. 123–213.

(pp. 96–97) A brief description of lactic acid production is given by R. Datta and S.-P. Tsai, in "Fuels and Chemicals from Biomass," ed. B. C. Saha and J. Woodward, American Chemical Society Symposium Series 666 (American Chemical Society, Washington, D.C., 1997), pp. 224–236.

Lactic acid production and polymerization of lactic acid are reviewed by M. H. Hartmann, in *Biopolymers from Renewable Resources* (1998), *op. cit.*, pp. 367–411.

Ring-opening polymerizations are described by P. Dubois, P. Degée, N. Ropson, and R. Jérôme, in *Macromolecular Design of Polymeric Materials*, ed. K. Hatada, T. Kitayama, and O. Vogl (Marcel Dekker, New York, 1997), pp. 247–272.

(p. 97) Commercial poly(aspartic acid) and its uses are described by K. C. Low, A. P. Wheeler, and L. P. Koskan, in *Hydrophilic Polymers: Performance with Environmental Acceptability* (1996), *op. cit.*, pp. 99–111. Its biodegradability is described by M. B. Freeman, Y. H. Paik, G. Swift, R. Wilczynski,

S. K. Wolk, and K. M. Yocom, in *Hydrogels and Biodegradable Polymers for Bioapplications* (1996), *op. cit.*, pp. 118–136.

(pp. 97–100) For more on plant oils, see accounts by T. A. McKeon, J.-T. Lin, M. Goodrich-Tanrikulu, and A. Stafford, in *Agricultural Materials as Renewable Resources: Nonfood and Industrial Applications* (1996), *op. cit.*, pp. 158–178; M. O. Bagby, ibid., pp. 248–257; and S. F. Thames, M. D. Blanton, S. Mendon, R. Subramanian, and H. Yu, in *Biopolymers from Renewable Resources* (1998), *op. cit.*, pp. 249–280.

The castor oil statistics are from *Biobased Industrial Products* (2000), *op. cit.*, p. 50.

(p. 100) Silk is described by D. L. Kaplan, S. J. Lombardi, W. S. Muller, and S. A. Fossey, in *Biomaterials: Novel Materials from Biological Sources* (1991), *op. cit.*, pp. 1–53; D. L. Kaplan, C. M. Mello, S. Arcidiacono, S. A. Fossey, K. Senecal, and W. S. Muller, in *Protein-Based Materials* (1997), *op. cit.*, pp. 103–131; S. M. Hudson, ibid., pp. 313–337; and J. P. Anderson, ibid., pp. 371–423. See also *Silk Polymers: Materials Science and Biotechnology*, ed. D. L. Kaplan, E. W. Adams, B. Farmer, and C. Venèy, American Chemical Society Symposium Series 544 (American Chemical Society, Washington, D.C., 1994).

(p. 101) Silk and mussel fibers are described in *Biopolymers: Making Materials Nature's Way* (1993), *op. cit.*, pp. 26–27.

The block copolymer nature of the mussel protein is described by K. J. Coyne, X.-X. Qin, and J. H. Waite, *Science* 277 (19 September 1997): 1830–1832.

(pp. 101–103) Nature's use of biopolymers in constructing the "materials" of living organisms is well described in *Made to Measure: New Materials for the 21st Century*, Chapter 4, by P. Ball (Princeton University Press, Princeton, 1997).

Chapter 7 The Reemergence of Bioplastics

(pp. 105–113) The historical commentary is from *Pioneer Plastic: The Making and Selling of Celluloid*, by R. Friedel (University of Wisconsin Press, Madison, 1983); C. J. Williamson, in *The Development of Plastics*, ed. S.T.I. Mossman and P.J.T. Morris (Royal Society of Chemistry, Cambridge, 1994), pp. 1–9; S.T.I. Mossman, ibid., pp. 10–25; *Plastics Materials*, 7th edition, by J. A. Brydson (Butterworth-Heinemann, Oxford, 1999), pp. 1–6, 831–838; *American Plastic: A Cultural History*, by J. L. Meikle (Rutgers University Press, New Brunswick, New Jersey, 1995); *Plastic: The Making of a Synthetic Century*, by S. Fenichell (HarperCollins, New York, 1996); *Celluloid: Collector's Reference and Value Guide*, by K. Lauer and J. P. Robinson (Collector Books, Paducah, Kentucky, 1999); and *The Story of Casein—from Milk Paint to Plastic*, by J. P. Robinson, *Antique Week*, 14 February 2000.

(pp. 105–106) Examples of early Native American use of horn as a processed material are found in *Native Visions*, by S. C. Brown, with photographs by P.

Macapio (Seattle Art Museum, in association with the University of Washington Press, Seattle, 1998).

(pp. 113–115) The account of Henry Ford's use of soy plastics is from E. F. Lougee, "Industry and the Soy Bean," in *Modern Plastics*, April 1936, p. 13; *Time* magazine, 11 November 1940, p. 65; "Ford Builds a Plastic Auto Body," in *Modern Plastics*, September 1941; P. S. Sprague, "Utilization of Soya Protein in Industry," *Soybean Digest*, September 1944, pp. 47–48, 51; R. Davis, "Henry's Plastic Car: An Interview with Mr. Lowell E. Overly," Ford Museum Library, Dearborn, Michigan, vertical file: Plastics–Soybeans–Automobiles; and Raymond W. Wik, *Henry Ford and Grassroots America*, (University of Michigan Press, Ann Arbor, 1972), pp. 149–151. See also J. L. Meikle, *American Plastic: A Cultural History*, (Rutgers University Press, New Brunswick, New Jersey, 1995), pp. 155–157; and S. Fenichell, *Plastic: The Making of a Synthetic Century*, (HarperCollins, New York, 1996), pp. 175–180.

(pp. 116–117) The description of cellophane is taken from F. Reiter, *Packaging* 31 (1986): 52–53; and from J. M. Krochta and C.L.C. DeMulder-Johnston, in *Agricultural Materials as Renewable Resources: Nonfood and Industrial Applications* (1996), *op. cit.*, pp. 120–140.

(p. 118) The thermoplastic behavior of starch is reviewed by S. Simmons, C. E. Weigand, R. J. Albalak, R. C. Armstrong, and E. L. Thomas, in *Biodegradable Polymers and Packaging* (1993), *op. cit.*, pp. 171–207; G. M. Chapman, in *Polymers from Agricultural Coproducts* (1994), *op. cit.*, pp. 29–47; and C. Bastioli, in *Degradable Polymers: Principles and Applications* (1995), *op. cit.*, pp. 112–137.

(pp. 118–123) Starch-based plastics are described by W. M. Doane, C. L. Swanson, and G. F. Fanta, in *Emerging Technologies for Materials and Chemicals from Biomass* (1992), *op. cit.*, pp. 197–230; R. L. Shogren, G. F. Fanta, and W. M. Doane, in "Development of Starch-Based Plastics—A Reexamination of Selected Polymer Systems in Historical Perspectives," *Starch/Stärke* 45 (1993): 276–280; G.J.L. Griffin, in *Chemistry and Technology of Biodegradable Plastics* (1994), *op. cit.*, pp. 18–47, 135–150; and R. L. Shogren, in *Biopolymers from Renewable Resources* (1998), *op. cit.*, pp. 30–46.

(p. 119) For studies of the degradability of starch-polyethylene plastics, see, for example, the works of A.-C. Albertsson and coworkers C. Barenstedt, S. Karlsson, and T. Lundberg, *Polymer* 36 (1995): 3075–3083; and A.-C. Albertsson and S. Karlsson, in *Macromolecular Design of Polymeric Materials*, ed. K. Hatada, T. Kitayama, and O. Vogl (Marcel Dekker, New York, 1997), pp. 793–802. See also a long-term study by S. R. Yang and C. H. Wu, in *Macromolecular Symposia* 144 (1999): 101–112.

Starch-poly(vinyl alcohol) plastics are described by C. Bastioli, V. Bellotti, L. Del Giudice, and G. Gilli, in *Biodegradable Polymers and Plastics* (1992), *op. cit.*, pp. 101–111. (See fig. 7.6.)

(p. 120) Thermoplastic blends of starch and polycaprolactone are described by L. Avérous, L. Moro, P. Dole, and C. Fringant, in *Polymer* 41 (2000): 4157–4167.

For a description of thermoplastic starch sheets laminated with a water-resistant polycaprolactone coating, see "Preparation and Properties of Thermoplastic Starch-Polyester Laminate Sheets by Coextrusion," by L. Wang, R. L. Shogren, and C. Carriere, *Polymer Engineering Science* 40 (2000): 499–506. (See fig. 7.7.)

(pp. 121–122) Starch-based foams are described by G. M. Glenn, R. E. Miller, and D. W. Irving, in *Agricultural Materials as Renewable Resources: Non-food and Industrial Applications* (1996), *op. cit.*, pp. 88–106; and by G. M. Glenn, W. J. Orts, R. Buttery, and D. Stern, in *Polymers from Renewable Resources: Polysaccharides and Agroproteins* (2001), *op. cit.*, pp. 42–60.

Starch foamed articles for packaging and serving needs are reported by R. L. Shogren, J. W. Lawton, W. M. Doane, and R. F. Tiefenbacher, in "Structure and Morphology of Baked Starch Foams," *Polymer* 39 (1998): 6649–6655; and by R. L. Shogren, J. W. Lawton, K. F. Tiefenbacher, and L. Chen, in "Starch-poly(vinyl alcohol) Foamed Articles Prepared by a Baking Process," *Journal of Applied Polymer Science* 68 (1998): 2129–2140. (See fig. 7.8.)

Water-resistant foam peanuts made from starch acetate, with a 2.5 degree of acetate substitution, and water are described by R. L. Shogren, in "Preparation, Thermal Properties, and Extrusion of High-Amylose Starch Acetates," *Carbohydrate Polymers* 29 (1995): 57–62 (fig. 7.9). See also "Biodegradable Starch Based Coating to Waterproof Hydrophilic Materials," by C. Fringant, M. Rinaudo, N. Gontard, S. Guilbert, and H. Derradji, *Starch-Stärke* 50 (1998): 292–296.

For an example of the effects of chemical modification on biodegradability, see work by D. S. Roessner, S. P. McCarthy, R. A. Gross, and D. L. Kaplan, *Macromolecules* 29 (1996): 1–6.

(pp. 123–124) For reviews of chitin and chitosan as fibers and film formers, see articles by T. D. Rathke and S. M. Hudson, *Journal of Macromolecular Science, Reviews of Macromolecular Chemistry and Physics* C34 (1994): 375–437; and by S. M. Hudson and C. Smith, in *Biopolymers from Renewable Resources* (1998), *op. cit.*, pp. 96–118.

References to wound-healing applications of chitosan, and chitin, are given by S. Minami, Y. Okamoto, S.-I. Tanioka, H. Sashiwa, H. Saimoto, A. Matsuhashi, and Y. Shigemasa, in *Carbohydrates and Carbohydrate Polymers* (1993), *op. cit.*, pp. 141–152; and by S.-I. Tanioka, Y. Okamoto, S. Minami, A. Matsuhashi, S. Tokura, H. Sashiwa, H. Saimoto, and Y. Shigemasa, ibid., pp. 153–164.

Fiber blends are described by S. Hirano and M. Zhang, *Carbohydrate Polymers* 43 (2000): 281–284.

For a study of the effect of chitosan film acylation on biodegradability, see a report by J. Xu, S. P. McCarthy, R. A. Gross, and D. L. Kaplan, *Macromolecules* 29 (1996): 3436–3440.

Studies of chitosan-poly(vinyl alcohol) blends plasticized with sorbitol and sucrose have been reported by I. Arvanitoyannis, I. Kolokuris, A. Nakayama, N. Yamamoto, and S. Aiba., *Carbohydrate Polymers* 34 (1997): 9–19.

Cellulose-chitosan films are described by M. Nishiyama, J. Hosokawa, K. Yoshihara, T. Kubo, H. Kabeya, T. Endo, and R. Kitagawa, in *Hydrophilic Polymers: Performance with Environmental Acceptability* (1996), *op. cit.*, pp. 113–123.

(p. 124) Starch-pectin films are described by D. R. Coffin and M. L. Fishman, in *Polymers from Agricultural Coproducts* (1994), *op. cit.*, pp. 82–91.

The degradation of starch-calcium carbonate disposable packaging is described by V. T. Breslin, in *Journal of Environmental Polymer Degradation* 6 (1998): 9–21.

Structure-function aspects of proteins as materials are described in "Protein Composite Materials," by P. Calvert, in *Protein-Based Materials* (1997), *op. cit.*, pp. 180–216; and in "Protein-Based Materials," by M. M. Butler and K. P. McGrath, in *Biopolymers from Renewable Resources* (1998), *op. cit.*, pp. 177–194.

Soy protein plastics are described by Y. T.-P. Ly, L. A. Johnson, and J. Jane, in *Biopolymers from Renewable Resources* (1998), *op. cit.*, pp. 144–176; and by X. S. Sun, in *Polymers from Renewable Resources: Polysaccharides and Agroproteins* (2001), *op. cit.*, pp. 132–148.

(p. 125) Edible food-barrier coatings are reviewed by K. R. Conca and T.C.S. Yang, in *Biodegradable Polymers and Packaging* (1993), *op. cit.*, pp. 357–369; and by F. Debeaufort, J.-A. Quezado-Gallo, and A. Voilley, in *Food Packaging: Testing, Methods and Applications* (2000), *op. cit.*, pp. 9–16.

Edible films for produce are described by A. E. Pavlath, D.S.W. Wong, J. Hudson, and G. H. Robertson, in *Agricultural Materials as Renewable Resources: Nonfood and Industrial Applications* (1996), *op. cit.*, pp. 107–119.

Films and coatings from commodity agroproteins are described by N. Parris, L. C. Dickey, P. M. Tomasula, D. R. Coffin, and P. J. Vergano, in *Polymers from Renewable Resources: Polysaccharides and Agroproteins* (2001), *op. cit.*, pp. 118–131.

A study of edible films made from starch-caseinate blends is reported by I. Arvanitoyannis, E. Psomiadou, and A. Nakayama, *Carbohydrate Polymers* 34 (1997): 179–192.

Composite edible films are described by T. H. McHugh, in *Macromolecular Interactions in Food Technology* (1996), *op. cit.*, pp. 134–144; and by P. D. Hoagland, ibid., pp. 145–154.

For a study of the biodegradability of starch-protein plastics, see K. E. Spence, A. L. Allen, S. Wang, and J. Jane, in *Hydrogels and Biodegradable Polymers for Bioapplications* (1996), *op. cit.*, pp. 149–158.

Reference to starch-protein composites as meeting the nutritional requirements for farm animals is made by J.-L. Jane, S.-T. Lim, and I. Paetau, in *Biodegradable Polymers and Packaging* (1993), *op. cit.*, p. 64.

(pp. 125–128) The properties and applications of polyhydroxyalkanoates are described in *Microbial Polyesters*, by Y. Doi (Wiley, New York, 1990). See also reports by M. K. Cox, in *Biodegradable Polymers and Plastics* (1992), *op. cit.*, pp. 95–100; O. Hrabak, ibid., pp. 255–258; P. J. Hocking and R. H. Marchessault, in *Chemistry and Technology of Biodegradable Polymers* (1994), *op. cit.*, pp. 48–96; T. Hammond and J. J. Liggat, in *Degradable Polymers: Principles and Applications* (1995), *op. cit.*, pp. 88–112; P. J. Hocking and R. H. Marchessault, in *Biopolymers from Renewable Resources* (1998), *op. cit.*, pp. 220–248; and L.J.R. Foster, in *Polymers from Renewable Resources: Biopolyesters and Biocatalysis* (2000), *op. cit.*, pp. 42–66.

The processing of PHBV is reviewed by D. Kemmish, in *Biodegradable Polymers and Packaging* (1993), *op. cit.*, pp. 225–245.

For a description of enhancing the water resistance of starch foam products by laminating PHBV coatings with a shellac adhesive, see R. L. Shogren, "Water Vapor Permeability of Biodegradable Polymers," *Journal of Environmental Polymer Degradation* 5 (1997): 91–95; and R. L. Shogren and J. W. Lawton, "Enhanced Water Resistance of Starch-Based Materials," U.S. Pat. 5,756,194, 1998 (fig. 7.10).

For studies of the degradation of PHBV, see, for example, M. R. Timmins, D. F. Gilmore, R. C. Fuller, and R. W. Lenz, in *Biodegradable Polymers and Packaging* (1993), *op. cit.*, pp. 119–131; L.J.R. Foster, R. C. Fuller, and R. W. Lenz, in *Hydrogels and Biodegradable Polymers for Bioapplications* (1996), *op. cit.*, pp. 68–92; C. L. Yue, R. A. Gross, and S. P. McCarthy, *Polymer Degradation and Stability* 51 (1996): 205–210; R. Renstad, S. Karlsson, and A.-C. Albertsson, in *"Petro" Polymers vs. "Green" Polymers* (1998), *op. cit.*, pp. 241–249; and T. M. Scherer, M. M. Rothermich, R. Quinteros, M. T. Poch, R. W. Lenz, and S. Goodwin, in *Polymers from Renewable Resources: Biopolyesters and Biocatalysis* (2000), *op. cit.*, pp. 254–280.

Biomedical applications of PHBV are overviewed by C. Scholz, in *Polymers from Renewable Resources: Biopolyesters and Biocatalysis* (2000), *op. cit.*, pp. 328–334.

(pp. 128–130) A racemic mixture of L-lactic acid and D-lactic acid promotes an amorphous PLA with low melting point. PLA from L- or D-monomers are semicrystalline. A racemic catalyst, however, can promote the polymerization of a racemic mixture of monomers to yield a stereoregular complex with high crystallinity, and high melting point. Stereoselective polymerization of poly(lactic acid) is reported by C. P. Radano, G. L. Baker, and M. R. Smith in *Journal of the American Chemical Society* 122 (2000): 1552–1553.

The environmental degradation of poly(lactic acid) is described by S. Karlsson and A.-C. Albertsson, in *"Petro" Polymers vs. "Green" Polymers* (1998), *op. cit.*, pp. 219–225.

For the effects of crystallinity on the enzymatic degradation of poly(lactic acid), see work by H. Cai, V. Dave, R. A. Gross, and S. P. McCarthy, *Journal of Polymer Science*, Part B: *Polymer Physics* 34 (1996): 2701–2708.

For a report on blends of poly(lactic acid) and poly(ethylene glycol), see, for example, that of M. Sheth, R. A. Kumar, V. Dave, R. A. Gross, and S. P. McCarthy, *Journal of Applied Polymer Science* 66 (1997): 1495–1505.

The heparin delivery system is described by E. R. Edelman, A. Nathan, M. Katada, J. Gates, and M. J. Karnovsky, in *Biomaterials* 21 (2000): 2279–2286.

The 5-fluorouracil delivery system is described by T. Chandy, G. S. Das, and G.H.R. Rao, in *Journal of Microencapsulation* 17 (2000): 625–638.

The bone regeneration matrix is described by C. T. Laurencin, A.M.A. Ambrosio, M. A. Attawia, F. K. Ko, and M. D. Borden, in *Polymers from Renewable Resources: Biopolyesters and Biocatalysis* (2000), *op. cit.*, pp. 294–310.

The coated implant study is reported by G. Schmidmaier, B. Wildemann, H. Bail, M. Lucke, A. Stemberger, A. Flyvbjerg, and M. Raschke, in *Chirurg* 71 (2000): 1016–1022.

(pp. 131–132) For a description of fiber-reinforced triglyceride resins, see "Development of Affordable Soy-Based Plastics, Resins, and Adhesives," by R. P. Wool, in *CHEMTECH,* June 1999, 44–48 (fig. 7.11).

(pp. 132–133) Biodegradable mulch consisting of kraft paper coated with a polyester made from a reaction product of epoxidized soybean oil and citric acid is described by R. L. Shogren, *Polymer Preprints* 39 (1998): 91–92; *Journal of Applied Polymer Science* 73 (1999): 2159–2167; and *Journal of Sustainable Agriculture* 16 (2000): 33–47 (fig. 7.12).

(p. 133) A summary report on composting bags is given by K. Shaw, in *BioCycle* 40, 7 (July 1999): 75–82. A comparative performance study of the compostability of biodegradable bags is reported by R. E. Farrell, T. J. Adamczyk, D. C. Broe, J. S. Lee, B. L. Briggs, R. A. Gross, S. P. McCarthy, and S. Goodwin, in *Polymers from Renewable Resources: Polysaccharides and Agroproteins* (2001), *op. cit.*, pp. 337–375. See also a report by E. Epstein, ibid., pp. 387–398.

(p. 134) Starch-PHBV blends have been described by S. J. Huang, M. F. Koenig, and M. Huang, in *Biodegradable Polymers and Packaging* (1993), *op. cit.*, p. 106.

Chapter 8 Factors Affecting Growth

(p. 136) The data for figure 8.1 are only estimates, based on current chemical supply-house catalogs, for similar quantities of each substance.

The chitin biomass estimate is taken from S. Hirano, N. Hutadilok, K.-I. Hayashi, K. Hirochi, A. Usutani, and H. Tachibana, in *Carbohydrates and Carbohydrate Polymers* (1993), *op. cit.*, pp. 253–264.

(p. 138) The concern over phthalates in PVC is reported by B. Hileman, in *Chemical and Engineering News*, 13 December 1999 and 7 August 2000.

(p. 140) A brief survey of important bioplastics physical properties is given by J. M. Krochta and C.L.C. DeMulder-Johnston, in *Agricultural Materials as Renewable Resources: Nonfood and Industrial Applications* (1996), *op. cit.*, pp. 120–140.

(pp. 140–141) Chemical modification of natural materials is discussed by T. Ouchi and Y. Ohya, in *Macromolecular Design of Polymeric Materials*, ed. K. Hatada, T. Kitayama, and O. Vogl (Marcel Dekker, New York, 1997), pp. 351–364.

Native starch substituted with caprolactone is described by C. Pellegrini and I. Tomka, in *"Petro" Polymers vs. "Green" Polymers* (1998), *op. cit.*, pp. 31–35.

Protein modifications are described by L. A. de Graaf and P. Kolster, in *"Petro" Polymers vs. "Green" Polymers* (1998), *op. cit.*, pp. 51–58.

For an example of water resistance modifications applied to starch-zein mixtures, see S. T. Lim and J.-L. Jane, in *Carbohydrates and Carbohydrate Polymers* (1993), *op. cit.*, pp. 288–297.

Water-soluble laundry bags are described by W. M. Doane, C. L. Swanson, and G. F. Fanta, in *Emerging Technologies for Materials and Chemicals from Biomass* (1992), *op. cit.*, pp. 206–230.

(p. 142) For an example of tensile strength modification applied to chitosan films, see T. D. Rathke and S. M. Hudson, in *Carbohydrates and Carbohydrate Polymers* (1993), *op. cit.*, pp. 281–287.

A study of the varying crystallinity of cast starch films as a function of film formation conditions was reported by A. Rindlav, S.H.D. Hulleman, and P. Gatenholm, *Carbohydrate Polymers* 34 (1997): 25–30.

(p. 144) The NRC study panel report is cited in *Chemical & Engineering News*, 26 July 1999, p. 26.

The cost of wheat relative to oil and its change from 1967 to 1990 is from C. A. Spelman, in *Non-Food Uses of Agricultural Raw Materials: Economics, Biotechnology, and Politics* (1994), *op. cit.*, p. 92.

Chapter 9 Prospects for the Future

(pp. 145–146) The potential use of biomass plantations for nonfood applications is evaluated by R. Williams, in *Industrial Ecology and Global Change*, ed. R. Socolow, C. Andrews, F. Berkhout, and V. Thomas (Cambridge University Press, Cambridge, 1994), pp. 199–225. Cited statistics are from the same source. See also *Non-Food Uses of Agricultural Raw Materials: Economics, Biotechnology and Politics* (1994), *op. cit.*

(p. 147) The use of rapeseed oil methyl ester as a carbon source for the production of polyhydroxyalkanoates is described by A. Steinbüchel, I. Voss, and V. Gorenflo, in *Polymers from Renewable Resources: Biopolyesters and Biocatalysis* (2000), *op. cit.*, pp. 14–24.

The use of other plant oils and animal fats is described by R. D. Ashby, D.K.Y. Solaiman, and T. A. Foglia, ibid., pp. 25–41.

Projects aimed at using food industry wastes and municipal activated sludge to produce polyhydroxyalkanoates are described by P. H. Yu, H. Chua, and P.A.L. Huang, *Macromolecular Symposia* 148 (1999): 415–424; and by A. L. Wong, H. Chua, W. H. Lo, and P.H.F. Yu, *Water Science and Technology* 41 (2000): 55–59.

The cloning experiment was first performed by Douglas Dennis at James Madison University in Virginia. The often cited article "In Search of the Plastic Potato," by Robert Pool, appeared in *Science* 245 (15 September 1989): 1187–1189.

A brief report on transgenic plants, including those engineered to produce PHB and PHBV, appeared in *Science* 282 (18 December 1998): 2176–2178.

(pp. 147–148) For more on the conversion of lignocellulosics to fermentable sugars, see *Biobased Industrial Products* (2000), *op. cit.*

(p. 149) For a description of biomedical applications, see *Polymers in Medicine and Pharmacy*, Vol. 394, ed. A. G. Mikos, K. W. Leong, M. J. Yaszemski, J. A. Tamada, and M. L. Radomsky, Materials Research Society Symposium Proceedings (Materials Research Society, Pittsburgh, Pennsylvania, 1995).

(pp. 150–151) The role of biotechnology in expanding the raw material resource base is described in *Biobased Industrial Products* (2000), *op. cit.*, pp. 26–54.

Enzymatic catalysis of polymerizations is described in *Polymers from Renewable Resources: Biopolyesters and Biocatalysis* (2000), *op. cit.*

(p. 152) Strong environmentalist statements are found in *Making Peace with the Planet*, by B. Commoner (Pantheon Books, New York, 1990); *Betrayal of Science and Reason: How Anti-Environmental Rhetoric Threatens Our Future*, by P. R. Ehrlich and A. H. Ehrlich (Island Press, Washington, D.C., 1996); and *Water: The Fate of Our Most Precious Resource*, by M. de Villiers (Houghton Mifflin, New York, 2000).

For a review article on the global carbon cycle, see the report from a group of international scientific research programs, authored by P. Falkowski, R. J. Scholes, E. Boyle, J. Canadell, D. Canfield, J. Elser, N. Gruber, K. Hibbard, P. Högberg, S. Linder, F. T. Mackenzie, B. Moore III, T. Pedersen, Y. Rosenthal, S. Seitzinger, V. Smetacek, and W. Steffen, *Science* 290 (13 October 2000): 291–296.

For a description of a "cradle-to-grave" analysis of the environmental impacts of polymers, see B. W. Vignon, D. A. Tolle, and V. McGinnis, in *Ecological Assessment of Polymers: Strategies for Product Stewardship and Regulatory Programs* (1997), *op. cit.*, pp. 307–335.

(p. 153) Reference to the MARPOL treaty is taken from *Biodegradable Polymers and Packaging* (1993), *op. cit.*, p. xiv.

The Navy's interest in biodegradable materials for military applications is described by J. M. Ross, in *Biodegradable Polymers and Packaging* (1993), *op. cit.*, pp. 381–388.

(p. 154) Germany's Green Dot program is summarized by F. Mader, in *Chemical Aspects of Plastics Recycling* (1997), *op. cit.*, pp. 13–27; and noted in *Chemical & Engineering News*, 7 July 1997, p. 23. A more detailed description is given by B. Goeke and F. Chalot, in *Washington Waste Minimisation Workshop*, Vol. 1 (Organisation for Economic Co-operation and Development, Paris, 1996), pp. 11–102. The recovery statistics are from *BioCycle* 40, 8 (August 1999): 12, and the gruener-punkt.de website.

(pp. 156–157) For an analysis of industry's role in sustaining the environment, and the inclusion of environmental costs and benefits in corporate decision making, see *Eco-Efficiency: The Business Link to Sustainable Development*, by L. D. DeSimone and F. Popoff (MIT Press, Cambridge, 1997). See also the review of that work by J. R. Hirl, *Chemical & Engineering News*, 13 April 1998, pp. 50–51.

The impact of the Responsible Care program is from E. W. Deavenport Jr., cited in *Chemical & Engineering News*, 5 October 1998, p. 26. See also "Responsible Care," by M. S. Reisch, in *Chemical & Engineering News*, 4 September 2000, pp. 21–26.

Reports on sustainable development within the chemical industry appeared in *Chemical & Engineering News*, 6 December 1999, pp. 55–70; and 3 September 2001, pp. 17–22.

(p.157) IBM became the first major multinational company to obtain a worldwide ISO 14001 registration, in late 1997, illustrating the private sector's leadership potential in achieving sustainable development.

(pp. 157–158) The first article on industrial ecology written for a general audience was perhaps "Strategies for Manufacturing," by R. A. Frosch and N. E. Gallopoulos, which appeared in a special issue of *Scientific American* entitled "Managing Planet Earth," 261 (September 1989): 144–153. The authors referred to an "industrial ecosystem" as an analog of biological ecosystems. R. A. Frosch, then Vice President and Director of Research at General Motors, and a member of both the United States National Academy of Sciences (NAS) and National Academy of Engineering (NAE), worked to promote the concept of industrial ecology, and initiated coalescing events such as a 1991 NAS colloquium and succeeding NAE workshops.

For more on industrial ecology, see *Industrial Ecology*, by B. R. Allenby and T. E. Graedel (Prentice-Hall, Upper Saddle River, New Jersey, 1995); *Industrial Ecology: Towards Closing the Materials Cycle*, Chapters 1, 2, 12, 15, and 16, by R. U. Ayres and L. W. Ayres (Edward Elgar Publishing, Cheltenham, U.K., 1996); *Industrial Ecology and Global Change*, ed. R. H. Socolow, C. Andrews, F. Berkhout, and V. Thomas (Cambridge University

Press, Cambridge, 1997); and *Vers une écologie industrielle*, by S. Erkman (Éditions Charles Léopold Mayer, Paris, 1998).

A brief article on industrial ecology, by B. Hileman, appeared in *Chemical & Engineering News*, 20 July 1998, pp. 41–42. See also response letters, ibid., 10 August 1998, pp. 6–7.

(p. 160) The Chemical Congress program is from *Chemical & Engineering News*, 15 September 1997, pp. 56–57.

Chrysler's Composite Concept Vehicle was noted in *Chemical & Engineering News*, 15 September 1997, p. 11.

The Cargill Dow Polymer LLC venture is described in *Chemical & Engineering News*, 1 December 1997, p. 7; 8 December 1997, pp. 14–16; and 17 January 2000, p. 13.

(p. 161) Dr. Paul S. Anderson's remarks are cited in *Chemical & Engineering News*, 12 January 1998, p. 161.

The analyst is Sano Shimoda, a biotechnology specialist, quoted in *Chemical & Engineering News*, 10 August 1998, p. 34.

The program for the American Chemical Society's Boston meeting appeared in *Chemical & Engineering News*, 27 July 1998. Many of the papers were subsequently published in *Polymers from Renewable Resources: Biopolyesters and Biocatalysis* (2000), *op. cit.*; and *Polymers from Renewable Resources: Polysaccharides and Agroproteins* (2001), *op. cit.*

The materials innovation awards were noted in *Chemical & Engineering News*, 27 July 1998.

(p. 162) See a brief report on the "Technology Vision 2020" agenda by M. S. Reisch, in *Chemical & Engineering News*, 9 August 1999, pp. 10–12.

The program for the American Chemical Society's San Francisco meeting apeared in *Chemical & Engineering News*, 28 February 2000.

(p. 163) The formation of the Japanese Organic Recycling Association was reported in *BioCycle* 41, 10 (October 2000): 62–63.

Appendix Make Your Own

(p. 165) A story on the Virginia high school students' serendipitous discovery appeared in the *Washington Post*, 26 July 1997.

(pp. 166–167) Studies of the role of glycerol and sorbitol in starch films are reported by D. Lourdin, H. Bizot, and P. Colonna, *Journal of Applied Polymer Science* 63 (1997): 1047–1053; and *Macromolecular Symposia* 114 (1997): 179–185; S. Gaudin, D. Lourdin, D. LeBotlan, J. L. Ilari, and P. Colonna, *Journal of Cereal Science* 29 (1999): 273–284; S. Gaudin, D. Lourdin, D. LeBotlan, P. Forssell, J. L. Ilari, and P. Colonna, *Macromolecular Symposia* 138 (1999): 245–248; and S. Gaudin, D. Lourdin, P. M. Forssell, and P. Colonna, *Journal of Carbohydrate Polymers* 43 (2000): 33–37.

(p. 170) The term *formulation*, or even *recipe*, is used here for simplicity, rather than the industry specific term *compound.*

"Bioglass" is also the name given to a ceramic bone-replacement material to which bone and natural tissue will bind, developed by Larry Hench and colleagues at the University of Florida at Gainesville. See, for example, *Made to Measure: New Materials for the 21st Century*, by P. Ball (Princeton University Press, Princeton, 1997), pp. 224–225.

(p. 176) Starch-glycerol films formed from evaporation of solvent have been known for a long time. They are described, with references, by R. L. Shogren, in *Biodegradable Polymers and Packaging* (1993), *op. cit.*, pp. 141–150. An early study of the preparation of cast starch and starch-glycerol films was reported by I. A. Wolff, H. A. Davis, J. E. Clusky, L. J. Gundrum, and C. E. Rist, *Industrial Engineering and Chemistry* 43 (1951): 915–919. U.S. Patent 2,822,581 (1958) is entitled "Amylose Films," by J. Muetgeert and P. Hiemstra. U.S. Patent 3,344,216 (1967) is entitled "Process of Casting Amylose Films," by T. F. Protzman, J. A. Wagner, and A. H. Young. Cast starch–poly(vinyl alcohol) films are described by R. P. Westhoff, W. F. Kwolek, and F. H. Otey, *Starch/Stärke* 31 (1979): 163–165.

(pp. 181–185) For more details on glass temperature, see, for example, *An Introduction to Polymer Science*, by H.-G. Elias (Wiley-VCH, Weinheim, Germany, 1997), pp. 315–320; for more on tensile testing, see ibid., pp. 329, 356, and 362. A thorough account of plastics testing is given by V. Shah, in *Handbook of Plastics Testing Technology* (Wiley, New York, 1984).

Glossary

A superscript one ([1]) indicates an ASTM definition. Definitions with a superscript two ([2]) are from the United States Environmental Protection Agency.

abiotic degradation: degradation not involving biological processes. (See also **biotic degradation**.)

addition polymerization: polymerization in which monomers are linked together without the splitting off of water or other simple molecules.[1] The original monomers usually contain carbon-carbon double bonds.

aerobic degradation: degradation of organic matter that takes place in the presence of oxygen, producing carbon dioxide. (See also **anaerobic degradation**.)

alga (pl., algae): any of a group of microscopic plants, chiefly aquatic, that are photosynthetic (i.e., that contain chlorophyll).

amino resin: a resin made by polycondensation of a compound containing amino groups, such as urea or melamine, with an aldehyde, such as formaldehyde, or an aldehyde-yielding material.[1]

anaerobic degradation: degradation of organic matter that takes place in the absence of oxygen, producing methane. (See also **aerobic degradation**.)

ASTM: American Society for Testing and Materials.

atomic weight: the weight of an atom according to a scale of atomic weight units valued as one-twelfth the mass of the ^{12}C atom, with the mass of a ^{12}C atom equal to 12.00 units exactly. One unit is equal to 1.6605×10^{-24} grams.

bacterium (pl., bacteria): any of a large group of microorganisms, usually consisting of a single cell; typical shapes are spherical, rodlike, spiral, or filamentous.

billion: in this book, a billion equals one thousand million (10^9).

biobased industrial products: fuels, chemicals, building materials, or electric power or heat produced from biomass. (U.S. Biomass Research and Development Act of 2000)

biochemical oxygen demand (BOD): the amount of oxygen consumed by microorganisms (mainly bacteria) and by chemical reactions in the biodegradation of organic matter.[2]

biodegradable: capable of being broken down into simpler compounds by the action of naturally occurring microorganisms such as bacteria, fungi, and algae.

biodegradable plastic: a degradable plastic in which the degradation results from the action of naturally occurring microorganisms such as bacteria, fungi, and algae.[1]

biodegradation (short for **biotic degradation):** chemical degradation brought about by the action of naturally occurring microorganisms such as bacteria, fungi, and algae.

biogeochemical cycle: the complex biological, geological, and chemical processes by which an atom or molecule is transformed within the ecosphere.

biomass: (1) the total mass of matter generated by the growth of living organisms, including plants, animals, and microorganisms; (2) any organic matter; (3) all of the living material in a given area; (4) often refers to vegetation; (5) any organic matter that is available on a renewable or recurring basis, including agricultural crops and trees, wood and wood wastes and residues, plants (including aquatic plants), grasses, residues, fibers, and animal wastes, municipal wastes, and other waste materials. (U.S. Biomass Research and Development Act of 2000)

bioplastics: in this book, biodegradable plastics whose components are derived entirely or almost entirely from renewable raw materials.

biopolymer (short for **biological polymer):** a polymer produced by plants, animals, or microorganisms through biochemical reactions, as distinct from a synthetic polymer.

biotechnology: broadly defined, biotechnology includes any technology that uses living organisms (or substances from those organisms) to make or modify a product, to improve plants or animals, or to develop microorganisms for specific uses. (U.S. Office of Technology Assessment, 1989)

biotic degradation: degradation involving biological processes. (See also **abiotic degradation**.)

blow molding: a method of fabrication in which a heated *parison* is forced into the shape of a mold cavity by internal gas pressure. A *parison* is the shaped plastic mass, generally in the form of a tube, used in blow molding.

BPS: Biodegradable Plastics Society (Japan).

carbon cycle: the complex series of chemical reactions by which carbon atoms are continually recycled within the ecosphere from one molecule to another.

cast film: a film made by depositing a layer of plastic, either molten, in solution, or in a dispersion, onto a surface, solidifying and removing the film from the surface.[1]

catalyst: a substance that accelerates a chemical reaction but undergoes no net chemical change.

cellular plastic: a plastic containing numerous cells, intentionally introduced, interconnecting or not, distributed throughout the mass.[1]

cellulosic plastics: plastics based on cellulose compounds, such as esters (cellulose acetate) and ethers (ethyl cellulose).[1]

CEN: Comité Européen de Normalisation (European Committee for Standardization).

chalking: (plastics) a powdery residue on the surface of a material resulting from degradation or migration of an ingredient, or both.[1]

chemical compound: a distinct and pure substance formed by the union of two or more elements in a definite proportion by weight.

composite: *n*—a solid product consisting of two or more distinct phases, including a binding material (matrix) and a particulate or fibrous material.[1] (Here, the term *phase* is meant in the general, descriptive sense, not the technical thermodynamic sense.) A broader definition includes not only heterogeneous but also homogeneous materials, such as polymer blends.

compost: (1) the product of **composting**[1] (q.v.); (2) the relatively stable humus material that is produced from a composting process in which bacteria in soil mixed with garbage and degradable trash breaks down the mixture into organic fertilizer;[2] (3) the humuslike, relatively **stabilized** material resulting from the biological decomposition or breakdown of organic materials.

compostable plastic: a plastic that undergoes biological degradation during composting to yield carbon dioxide, water, inorganic compounds, and biomass at a rate consistent with other known compostable materials and leaves no visually distinguishable or toxic residues.[1]

composting: (1) a managed process that controls the biological decomposition and transformation of biodegradable material into a humuslike substance called compost; the aerobic **mesophilic** and **thermophilic** degradation of organic matter to make compost; the transformation of biologically decomposable material through a controlled process of bio-oxidation that proceeds through mesophilic and thermophilic phases and results in the production of carbon dioxide, water, minerals, and **stabilized** organic matter (compost or humus). Composting uses a natural process to stabilize mixed decomposable organic material recovered from municipal solid waste, yard trimmings, biosolids (digested sewage sludge), certain industrial residues, and commercial residues;[1] (2) the controlled biological decomposition of organic material in the presence of air to form a humuslike material. Controlled methods of composting include mechanical mixing and aerating, ventilating the materials by dropping them through a vertical series of aerated chambers, or placing the compost in piles out in the open air and mixing it or turning it periodically.[2]

compression molding: the method of molding a material already in a confined cavity by applying pressure and usually heat.[1]

condensation polymerization: polymerization in which monomers are linked together with the splitting off of water or other simple molecules.[1] The monomeric repeating unit in the polymer lacks certain atoms that are present in the original monomer.

copolymer: a polymer formed from more than one type of monomer.

cress seed test: described in ASTM Standard Guide D 6002-96. Briefly, soil or compost is extracted with water and filtered. The supernatant is used for the germination test. Various dilutions of the supernatant are prepared, and aliquots are added to petri dishes lined with filter paper. Cress seeds are placed on the wet paper and left to germinate in the dark over four days at room temperature. The percentage of germinated seeds is determined after four days and compared with a water control. Strict adherence to the procedure requires that the samples be previously treated according to ASTM Practices 5152-91 or 5951-96.

cross-linking: the formation of a three-dimensional polymer by means of interchain reactions, resulting in changes in physical properties.[1]

cure: to change the properties of a polymeric system into a more stable, usable condition by the use of heat, radiation, or reaction with chemical additives;[1] cure may be accomplished, for example, by removal of solvent or by cross-linking (ISO).[1]

decomposition: the breakdown of matter by bacteria and fungi, changing the chemical makeup and physical appearance of materials.[2] (See also **degradation** and **deterioration**.)

degradable plastic: a plastic designed to undergo a significant change in its chemical structure under specific environmental conditions, resulting in a loss of some properties that may be measured by standard methods appropriate to the plastic and the application in a period of time that determines its classification.[1]

degradation: a deleterious change in the chemical structure, physical properties, or appearance of a plastic.[1] A change in the chemical structure of a plastic involving a deleterious change in properties (ISO). (See also **decomposition** and **deterioration**.)

deterioration: a permanent change in the physical properties of a plastic evidenced by impairment of those properties (ISO). (See also **decomposition** and **degradation**.)

DIN: Deutsches Institut für Normung (German Institute for Standardization).

dioxin: any of a family of compounds known chemically as dibenzo-*p*-dioxins.

earthworm test: described in ASTM Standard Guide D 6002-96. Briefly, compost containing the plastic material is mixed with a specified soil. The earthworm weight change and survival are measured. The results from compost containing plastic material are compared to controls of compost not containing the plastic material, and to controls of soil alone. Strict adherence to the procedure requires that the samples be previously treated according to ASTM Practices 5152-91 or 5951-96.

ecology: the interrelationship of animals, plants, and microorganisms with each other and the nonliving portions of their environment; also the study of that interrelationship.

ecosystem: the interacting system of a biological community and its nonliving environmental surroundings.[2]

elastomer: a macromolecular material that at room temperature returns rapidly to approximately its initial dimensions and shape after substantial deformation by a weak stress and release of the stress.[1]

element: in chemistry, a substance that cannot be separated into its constituent parts without losing its chemical identity; a fundamental substance comprising one kind of atom.

elongation: in tensile testing, the elongation of a specimen is the increase in length after rupture, referred to the original length. It is reported as percentage elongation.

emulsifier: a chemical that aids in suspending one liquid in another, usually an organic chemical in an aqueous solution; a substance that enables the mixing of normally unmixable liquids.

entropy: a measure of the disorder in a system.

environmental fate: the destiny of a chemical or biological pollutant after release into the environment.[2]

enzyme: a protein functioning as a biochemical catalyst.

extrusion: a process in which heated or unheated plastic is forced through a shaping orifice (a die) in one continuously formed shape, as in film, sheet, rod, or tubing.[1]

fabricating: the manufacture of plastic products from molded parts, rods, tubes, sheeting, extrusions, or other forms by appropriate operations such as punching, cutting, drilling, and tapping, including fastening plastic parts together or to other parts by mechanical devices, adhesives, heat sealing, or other means.[1]

feedstock: the raw materials used in an industry; e.g., petroleum is a feedstock for synthetic plastics, whereas starch, cellulose, soybeans, etc., are feedstocks for bioplastics.

fiber: the result of processing a polymer into a threadlike product, as distinct from a plastic or elastomer (although the distinction is not always sharp); any tough, threadlike substance.

filler: a relatively inert material added to a plastic to modify its strength, permanence, working properties, or other qualities, or to lower costs.[1]

film: in plastics, an optional term for sheeting having a nominal thickness not greater than 0.25 mm (0.01 in.).[1]

fossil fuels: fuels derived from ancient organic remains, e.g., peat, petroleum (crude oil), natural gas, and coal.

friable: easily crumbled, pulverized, or reduced to powder.

fungus (pl., **fungi**)**:** any of a group of microscopic plants that are aerobic and nonphotosynthetic (i.e., lack chlorophyll), including molds, mildews, mushrooms, yeasts, and others.

garbage: (1) solid waste that is properly placed in containers and becomes part of some waste management program, as distinct from **litter** that does not; (2) often used to distinguish wet, putrescent waste (garbage) from dry waste (**trash**), with the two collectively referred to as **refuse**;[2] (3) food waste resulting from the handling, storage, packaging, sale, preparation, cooking, and serving of foods.[2] (In the latter usage, not synonymous with **solid waste**, **trash**, or **rubbish**.)

genetic engineering: a process of inserting new genetic information into existing cells in order to modify a specific organism for the purpose of changing one or more of its characteristics.

glass temperature, or **glass transition temperature** (T_g)**:** the approximate midpoint of the temperature range over which the glass transition takes place.[1]

glass transition: the reversible change in an amorphous polymer or in amorphous regions of a partially crystalline polymer from (or to) a viscous or rubbery condition to (or from) a hard and relatively brittle one.[1]

greenhouse effect: a mechanism of warming Earth's atmosphere and surface by the absorption and re-radiation of long-wavelength (infrared) radiation by some atmospheric gases.

heavy metal: a metallic element with high atomic weight (e.g., mercury, chromium, cadmium, arsenic, lead); can damage living things at low concentrations and tends to accumulate in the food chain.[2]

homopolymer: a polymer resulting from polymerization involving a single monomer.[1]

humus: the organic portion of soil remaining after prolonged microbial decomposition;[2] a dark substance resulting from the partial decay of plant and/or animal matter.

hydrolysis: a chemical reaction in which a compound is converted into another compound by taking up the elements of water.

hydrolytically degradable plastic: a degradable plastic in which the degradation results from **hydrolysis**.[1]

hydrophilic: having a strong affinity for water; characterized by strong interactions with water molecules; literally, "water loving." (See also **hydrophobic**.)

hydrophobic: having a strong aversion for water; characterized by weak interactions with water molecules; literally, "water hating." (See also **hydrophilic**.)

incineration: a treatment technology involving destruction of waste by controlled burning at high temperatures.[2]

industrial ecology: the science of sustainability, which follows materials and energy from their source, through their conversion to products, to final reintegration into the natural biogeochemical cycles.

injection molding: the process of forming a material by forcing it, in a fluid state and under pressure, through a runner system (sprue, runner, gate[s]) into the cavity of a closed mold.[1]

inorganic: (1) not composed of once-living materials (e.g., minerals); (2) compounds of elements other than carbon, with the exception of a few simple compounds of carbon, including carbon monoxide, carbon dioxide, carbonates and cyanides, which are generally considered to be inorganic. (See also **organic**.)

ISO: International Standards Organization.

kinetic energy: energy possessed by virtue of motion.

kinetics: chemical kinetics is the branch of chemistry that deals with reaction rates.

kraft paper: a relatively coarse paper with high strength characteristics, produced by a modified sulfate process; unbleached grades are used mainly for wrapping and packaging.

laminate: a product made by bonding together two or more layers of material or materials (ISO).[1]

landfill: a method for final disposal of **solid waste** on land; a facility in which solid waste from municipal and/or industrial sources is disposed.[2] A **sanitary landfill** is a disposal site for nonhazardous solid waste spread in layers, compacted to the smallest practical volume, and covered by material applied at the end of each day.[2] Under current regulations, landfills are required to have liners and leachate treatment systems to prevent contamination of ground water and surface waters.[2]

lignin plastics: plastics based on lignin resins (ISO).[1]

lignin resin: a resin made by heating lignin or by reaction of lignin with chemicals or resins, the lignin being in greatest amount by mass (ISO).[1]

litter: the highly visible portion of **solid waste** carelessly discarded outside the regular garbage disposal system;[2] solid waste that does not become part of some waste management program. Solid waste properly placed in containers is often referred to as **garbage, trash,** or **refuse** (q.v.).

macromolecule: a long or large molecule not necessarily having a regularly repeating structural unit.

macroorganism: a visibly large organism.

MARPOL Annex V: an international agreement that bans the dumping of nondegradable plastics at sea.

material: in a general, descriptive sense, any composition of matter that can be fabricated into useful products; in a more technical sense, a composition of matter complying with defined standards. In the first sense *wood* is a material; in the latter it is not.

materials science: the study of compositions of matter that have application in the fabrication of useful products.

mesophilic phase: the phase of composting that occurs between 20 °C and 45 °C (68 °F and 113 °F).[1] (See also **thermophilic phase**.)

microbially active: containing microorganisms, as distinct from sterile.

microorganisms: mainly bacteria, fungi, and algae, but including protozoa, viruses, and a number of other organisms that are microscopic in size; i.e., that can be seen individually only with the aid of a microscope.

mineralization: (1) the conversion of biomass to gases (like carbon dioxide, methane, and nitrogen compounds), water, salts and minerals, and residual biomass; (2) the microbial conversion of an element from an organic to an inorganic state; the release of inorganic chemicals from organic matter in the process of aerobic or anaerobic decay.[2] Mineralization is complete when all the biomass is consumed and all carbon is converted to carbon dioxide.

mold test: described in ASTM Standard D 1924-63. Briefly, test specimens are placed in or on a solid agar growth medium deficient only in carbon. Observed growth depends on the test fungi utilizing the specimen as a carbon source. After three weeks, specimens are assigned growth ratings of 0—no growth, 1—traces (less than 10 percent covered), 2—light growth (10–30 percent covered), 3—moderate growth (30–60 percent covered), and 4—heavy growth (60–100 percent covered). Strips of filter paper serve as controls to ensure fungal activity.

molecular weight: the sum of atomic weights of all the atoms in a molecule.

molecule: the smallest division of a compound that still retains or exhibits all of the properties of the substance.

monomer: a low-molecular-weight substance consisting of molecules capable of reacting with like or unlike molecules to form a polymer.[1]

municipal solid waste (MSW): common garbage or trash generated by industries, businesses, institutions, and homes;[2] solid waste that includes nonhazardous residential waste, commercial waste, and institutional waste. For example, it excludes agricultural waste, biosolids, mining waste, industrial process waste, construction and demolition (C&D) debris, energy production waste (e.g., ash), abandoned automobiles, and street sweepings. Municipal landfills sometimes accept both MSW and C&D debris.

nanoscale technology: technology aimed at synthesizing macroscopic objects and devices to atomic specifications.

oligomer: a sustance composed of only a few monomeric units repetitively linked to each other, such as a dimer, trimer, tetramer, etc., or their mixtures.[1]

organic: (1) composed of living or once-living matter; (2) composed of compounds that contain carbon combined with other elements, such as hydrogen, oxygen, and nitrogen, but excluding a few simple compounds of carbon, including carbon monoxide, carbon dioxide, carbonates, and cyanides, which are generally considered to be inorganic. (See also **inorganic**.)

oxidation: the addition of oxygen to, the removal of hydrogen from, or the removal of electrons from an element or compound. (See also **reduction**.)

oxidatively degradable plastic: a degradable plastic in which the degradation results from oxidation.[1]

petroleum: crude oil or any fraction thereof that is liquid under normal conditions of temperature and pressure. The term includes petroleum-based substances constituting a complex blend of hydrocarbons derived from crude oil through the process of separation, conversion, upgrading, and finishing, such as motor fuel, jet oil, lubricants, petroleum solvents, and used oil.[2] Sometimes used to include natural gas plant liquids and liquefied refinery gases, and sometimes meant to include both crude oil and natural gas.

PHA: polyhydroxyalkanoate.

PHB: polyhydroxybutyrate.

PHBV: poly(hydroxybutyrate-*co*-hydroxyvalerate).

photodegradable plastic: a degradable plastic in which the degradation results from the action of natural daylight.[1]

photodegradation: degradation that results from the action of light.

photosynthesis: a process in which organisms, with the aid of chlorophyll contained in them, convert carbon dioxide and inorganic substances into oxygen and additional plant material, such as carbohydrates, using sunlight for energy; used by all green plants to grow.

PLA: poly(lactic acid); polylactate; polylactide.

plasticizer: a substance incorporated in a material to increase its workability, flexibility, or distensibility;[1] a substance that adds pliability to other substances.

plastic(s): a material that contains as an essential ingredient one or more organic polymeric substances of large molecular weight, is solid in its finished state, and, at some stage in its manufacture or processing into finished articles, can be shaped by flow. Rubber, textiles, adhesives, and paint, which may in some cases meet this definition, are not considered plastics.[1]

polyester: a polymer in which the repeated structural unit in the chain is of the ester type (ISO).

polyether: a polymer in which the repeated structural unit in the chain is of the ether type (ISO).

polymer: a substance consisting of molecules characterized by the repetition (neglecting ends, branch junctions, and other minor irregularities) of one or more types of monomeric units.[1]

polymerization: a chemical reaction in which the molecules of monomers are linked together to form polymers.[1]

polyolefin: a polymer prepared by the polymerization of an olefin(s) as the sole monomer(s).[1]

postconsumer recycling: reuse of materials generated from residential and consumer waste.[2]

protein: a complex nitrogen-containing organic compound of high molecular weight made of amino acids; many proteins are enzymes.

pyrolysis: a waste-treatment process in which material is heated in the absence of oxygen, and possibly under pressure as well. The process drives off volatile components and leaves a char. It is sometimes called destructive distillation.

recalcitrant: unresponsive; therefore, unreactive, resistant to degradation, or nondegradable.

recycle/reuse: minimizing waste generation by recovering and reprocessing usable products that might otherwise become waste (i.e., recycling of aluminum cans, paper, bottles, etc.)[2]

recycling: reusing materials and objects in original or changed forms rather than discarding them as waste;[2] in general usage, the *separation* of materials, such as newspapers, aluminum (and other metals), glass, and plastics from the waste stream, and their *processing*, so that the materials can be used again.

reduction: in chemistry, the removal of oxygen from, the addition of hydrogen to, or the addition of electrons to an element or compound. (See also **oxidation**.)

refuse: sometimes meant to be interchangeable with **solid waste;**[2] an inclusive term usually referring to wet discarded matter (**garbage**) and dry discarded matter (**trash**); in common usage, also often meant to be interchangeable with **garbage, trash,** and **rubbish**.

reinforced plastic: a plastic with high-strength fillers imbedded in the composition, resulting in some mechanical properties superior to those of the base resin.[1]

release agent: a material added to a compound or applied to the mold cavity, or both, to reduce parts sticking to the mold.[1]

resin: a solid or pseudosolid organic material often of high molecular weight, which exhibits a tendency to flow when subjected to stress, usually has a softening or melting range, and usually fractures conchoidally.[1] In a broad sense, the term is used to designate any polymer that is a basic material for plastics.[1]

rubber: a material that is capable of recovering from large deformations quickly and forcibly, and can be, or already is, modified to a state in which it is essentially insoluble (but can swell) in boiling solvent, such as benzene, methylethylketone, and ethanol-toluene azeotrope.[1]

rubbish: solid waste, excluding food waste and ashes, from homes, institutions, and workplaces.[2]

sanitary landfill: See **landfill**.

sheeting: a form of plastic in which the thickness is very small in proportion to length and width and in which the plastic is present as a continuous phase throughout, with or without filler.[1] (See also **film**.)

soil test: described in ASTM Standard Test Method D 5988-96, "Standard Test Method for Determining Aerobic Biodegradation in Soil of Plastic Materials or Residual Plastic Materials after Composting." Briefly, a defined soil/sand/manure compost matrix, or a natural soil sample, provides the microorganisms. The metabolism of test materials produces carbon dioxide, which is trapped in alkali solution and quantitated by titration. Readily biodegradable materials can be screened in 30 to 60 days.

solid waste: as defined under the United States Resource Conservation and Recovery Act, any solid, semisolid, liquid, or contained gaseous materials discarded from industrial, commercial, mining, or agricultural operations and from community activities. Solid waste includes garbage, construction debris, commercial refuse, sludge from water supply or waste treatment plants, or air-pollution-control facilities, plus other discarded materials.[2]

source reduction: reducing the amount of materials entering the waste stream by redesigning products or patterns of production or consumption (e.g., using returnable beverage containers); synonymous with waste reduction; the design, manufacture, purchase, or use of materials (such as products and packaging) to reduce the amount or toxicity of garbage generated. Source reduction can help reduce waste disposal and handling charges because the costs of recycling, municipal composting, landfilling, and combustion are avoided. Source reduction conserves resources and reduces pollution.[2]

stabilized: the point at which microbial activity reaches a low and relatively constant level (used in connection with the microbial decomposition of biodegradable material, including compost).

stabilizer: a substance that helps other substances resist chemical change; one type of plastics additive.

stewardship: taking responsibility.

Sturm test: described in ASTM Standard Test Method D 5209, "Test Method for Determining the Aerobic Biodegradation of Plastic Materials in the Presence of Municipal Sewer Sludge." Briefly, the metabolism of test materials produces carbon dioxide, which is trapped in alkali solution and quantitated by titration. The test length is typically 30 days. A positive result, of 60 percent recovery of the theoretically expected amount of carbon dioxide, usually indicates that the material will also biodegrade in a composting environment.

surfactant: a substance that reduces the surface tension of liquids, commonly used in detergents.

sustainable: able to be continued by virtue of consuming only renewable resources; able to keep up, sustain, or endure.

sustainable development: development that meets the needs of the present without compromising the ability of future generations to meet their own needs. (United Nations' World Commission on Environment and Development)

tensile strength: a measure of the force, per unit area, required to break a test specimen when it is placed between two clamps and then drawn; "the greatest longitudinal stress a substance can bear without tearing asunder." (Webster's)

thermodynamics: chemical thermodynamics is the branch of chemistry that deals with the interconversions of internal energy, heat, and work.

thermophilic phase: the phase in the composting process that occurs between 45 °C and 75 °C (113 °F and 167 °F); it is associated with specific colonies of microorganisms that accomplish a high rate of decomposition.[1] (See also **mesophilic phase**.)

thermoplastic: *n*—a plastic that repeatedly can be softened by heating and hardened by cooling through a temperature range characteristic of the plastic, and that in the softened state can be shaped by flow into articles by molding or extrusion;[1] *adj*—capable of being repeatedly softened by heating and hardened by cooling through a temperature range characteristic of the plastic, and that in the softened state can be shaped by flow into articles by molding or extrusion for example. Thermoplastic applies to those materials whose change upon heating is substantially physical.[1]

thermoset: *n*—a plastic that, after having been cured by heat or other means, is substantially infusible and insoluble;[1] *adj*—pertaining to the state of a plastic in which it is substantially infusible.[1]

thermosetting: *adj*—capable of being changed into a substantially infusible or insoluble product when cured by heat or other means.[1]

ton: a unit of weight in the United States Customary System of Measurement, equal to 2,000 pounds. It is also called a short ton, or net ton, and is equal to 0.907 metric ton, where one metric ton is equal to 1,000 kilograms.

toxicity: the quality or degree of being poisonous or harmful to plant, animal, or human life.[2]

transfer molding: a method of forming articles by fusing a plastic material in a chamber and then forcing essentially the whole mass into a hot mold where it solidifies.[1]

trash: material considered worthless or offensive that is thrown away; generally defined as dry waste material, but in common usage it is a synonym for **garbage**, **rubbish**, or **refuse**.[2]

vacuum forming: a forming process in which a heated plastic sheet is drawn against the mold surface by evacuating the air between it and the mold.[1]

virgin plastic: a plastic material in the form of pellets, granules, powder, floc, or liquid that has not been subjected to use or processing other than that required for its initial manufacture.[1]

viscose: (1) a thick liquid formed during the regeneration of cellulose used, for example, in making rayon; (2) generally, a thick mixture of solvent with a high concentration of polymer.

vulcanization: an irreversible process during which a rubber compound, through a change in its chemical structure (for example, cross-linking), becomes less plastic and more resistant to swelling by organic liquids and during which elastic properties are conferred, improved, or extended over a greater range of temperature.

waste: (1) unwanted materials left over from a manufacturing process;[2] (2) refuse from places of human or animal habitation.[2] See also **garbage**, **refuse**, **rubbish**, **solid waste**, **trash**, **waste stream**, and **yard waste**.

waste reduction: using source reduction, recycling, or composting to prevent or reduce waste generation.[2]

waste stream: the total flow of solid waste from homes, businesses, institutions, and manufacturing plants that is recycled, burned, or disposed of in landfills; or segments thereof such as the "residential waste stream" or the "recyclable waste stream."[2] Typical breakdown categories include agricultural waste, biosolids, mining waste, industrial process waste, construction and demolition debris, municipal waste, energy production waste (e.g., ash), abandoned automobiles, street sweepings, and others.

yard waste: the part of solid waste composed of grass clippings, leaves, twigs, branches, and garden refuse.[2]

Reading List

Agricultural Materials as Renewable Resources: Nonfood and Industrial Applications, ed. G. Fuller, T. A. McKeon, and D. D. Bills. American Chemical Society Symposium Series 647. American Chemical Society, Washington, D.C., 1996.

Biobased Industrial Products: Priorities for Research and Commercialization, Committee on Biobased Industrial Products, Board on Biology, Commission on Life Sciences, National Research Council. National Academy Press, Washington, D.C., 2000.

Biodegradable Polymers and Packaging, ed. C. Ching, D. L. Kaplan, and E. L. Thomas. Technomic, Lancaster, Pennsylvania, 1993.

Biodegradable Polymers and Plastics, ed. M. Vert, J. Feijen, A.-C. Albertsson, G. Scott, and E. Chiellini. Royal Society of Chemistry, Cambridge, 1992.

"Biodegradable Polymers and Plastics in Japan: Research, Development, and Applications," R. W. Lenz. National Technical Information Service Report, U.S. Dept. of Commerce, Washington, D.C., 1995.

Biodegradable Polymers and Recycling, ed. A.-C. Albertsson and S. J. Huang. *Journal of Macromolecular Science, Pure and Applied Chemistry* A32, 4 (1995): 593–904.

Biomaterials: Novel Materials from Biological Sources, ed. D. Byrom. Stockton Press, New York, 1991.

Biopolymers from Renewable Resources, ed. D. L. Kaplan. Springer-Verlag, Berlin, 1998.

"Biopolymers: Making Materials Nature's Way," U.S. Congress, Office of Technological Assessment Background Paper. U.S. Government Printing Office, Washington, D.C., 1993.

Carbohydrates and Carbohydrate Polymers, ed. M. Yalpani. ATL Press, Mount Prospect, Illinois, 1993.

Chemical Aspects of Plastics Recycling, ed. W. Hoyle and D. R. Karsa. Royal Society of Chemistry, Cambridge, 1997.

Chemistry and Technology of Biodegradable Polymers, ed. G.J.L. Griffin. Blackie Academic & Professional, imprint of Chapman and Hall, Glasgow, Scotland, 1994.

Degradable Polymers: Principles and Applications, ed. G. Scott and D. Gilead. Chapman and Hall, London, 1995.

"Directions for Environmentally Biodegradable Polymer Research," G. Swift. *Accounts in Chemical Research* 26 (1993): 105–110.

READING LIST

Ecological Assessment of Polymers: Strategies for Product Stewardship and Regulatory Programs, ed. J. D. Hamilton and R. Sutcliffe. Van Nostrand Reinhold, New York, 1997.

Emerging Technologies for Materials and Chemicals from Biomass, ed. R. M. Rowell, T. P. Schultz, and R. Narayan. American Chemical Society Symposium Series 476. American Chemical Society, Washington, D.C., 1992.

Food Packaging: Testing Methods and Applications, ed. S. J. Risch. American Chemical Society Symposium Series 753. American Chemical Society, Washington, D.C., 2000.

Hydrogels and Biodegradable Polymers for Bioapplications, ed. R. M. Ottenbrite, S. J. Huang, and K. Park. American Chemical Society Symposium Series 627. American Chemical Society, Washington, D.C., 1996.

Hydrophilic Polymers: Performance with Environmental Acceptability, ed. J. E. Glass. Advances in Chemistry Series 248. American Chemical Society, Washington, D.C., 1996.

Introduction to Polysaccharide Technology, M. P. Tombs and S. E. Harding. Taylor & Francis, London, 1998.

Macromolecular Interactions in Food Technology, ed. N. Parris, A. Kato, L. K. Creamer, and J. Pearce. American Chemical Society Symposium Series 650. American Chemical Society, Washington, D.C., 1996.

"Moving towards Consensus on Degradable Plastics," D. Riggle. *BioCycle* 39, 3 (March 1998): 64–70.

Non-Food Uses of Agricultural Raw Materials: Economics, Biotechnology and Politics, C. A. Spelman. Center for Agricultural and Biosciences International, Wallingford, U.K., 1994.

"Petro" Polymers vs. "Green" Polymers, ed. L. A. Kleintjens. Macromolecular Symposia 127. Hüthig & WepfVerlag, Zug, Switzerland, 1998.

Plastics, Rubber and Paper Recycling: A Pragmatic Approach, ed. C. P. Rader, S. D. Baldwin, D. D. Cornell, G. D. Sadler, and R. F. Stockel. American Chemical Society Symposium Series 609. American Chemical Society, Washington, D.C., 1995.

Polymers and the Environment, G. Scott. Royal Society of Chemistry, Cambridge, 1999.

Polymers from Agricultural Coproducts, ed. M. L. Fishman, R. B. Friedman, and S. J. Huang. American Chemical Society Symposium Series 575. American Chemical Society, Washington, D.C., 1994.

Polymers from Renewable Resources: Polyesters and Biocatalysis, ed. C. Scholz and R. Gross. American Chemical Society Symposium Series 764. American Chemical Society, Washington, D.C., 2000.

Polymers from Renewable Resources: Polysaccharides and Agroproteins, ed. R. Gross and C. Scholz. American Chemical Society Symposium Series 786. American Chemical Society, Washington, D.C., 2001.

Protein-Based Materials, ed. K. P. McGrath and D. L. Kaplan. Birkhäuser, Boston, 1997.

Author/Name Index

Subject Index

Franklin Pierce College Library
00147041